LANDSCAPES OF INVESTIGATION

Landscapes of Investigation

Contributions to Critical Mathematics Education

Edited by Miriam Godoy Penteado and Ole Skovsmose

Contributions by Adriana Lima, Agustín Méndez, Amanda Moura, Ana Carolina Faustino, Arindam Bose, Bülent Avcı, Daniela Soares, Débora Souza-Carneiro, Denner Barros, Edyenis Frango, Fanny Rodríguez, Fátima Oliveira, Guilherme Gomes da Silva, Íria Gaviolli, Jeimy Suaréz, Lessandra Marcelly, Lucicleide Bezerra, Manuella Carrijo, Mario Sánchez Aguilar, Miriam Penteado, Ole Skovsmose, Paula Civiero, Rafaela da Silva, Raquel Milani, Reginaldo Britto, Rejane Julio, Renato Ribeiro and Yael Moreno

https://www.openbookpublishers.com

© 2022 Miriam Godoy Penteado and Ole Skovsmose. Copyright of individual chapters is maintained by the chapters' authors

This work is licensed under an Attribution-NonCommercial 4.0 International (CC BY-NC 4.0). This license allows you to share, copy, distribute and transmit the text; to adapt the text for non-commercial purposes of the text providing attribution is made to the authors (but not in any way that suggests that they endorse you or your use of the work). Attribution should include the following information:

Miriam Godoy Penteado and Ole Skovsmose (eds), *Landscapes of Investigation: Contributions to Critical Mathematics Education*. Cambridge, UK: Open Book Publishers, 2022, https://doi.org/10.11647/OBP.0316

Copyright and permissions for the reuse of many of the images included in this publication differ from the above. This information is provided in the captions and in the list of illustrations.

Every effort has been made to identify and contact copyright holders and any omission or error will be corrected if notification is made to the publisher.

All external links were active at the time of publication unless otherwise stated and have been archived via the Internet Archive Wayback Machine at https://archive.org/web

Digital material and resources associated with this volume are available at https://doi.org/10.11647/OBP.0316#resources

Volume 1 | Studies on Mathematics Education and Society Book Series | ISSN Print: 2755-2616 | ISSN Digital: 2755-2624

ISBN Paperback: 978-1-80064-821-0
ISBN Hardback: 978-1-80064-822-7
ISBN Digital (PDF): 978-1-80064-823-4
ISBN Digital ebook (EPUB): 978-1-80064-824-1
ISBN Digital ebook (AZW3): 978-1-80064-825-8
ISBN XML: 978-1-80064-826-5
ISBN HTML: 978-1-80064-827-2
DOI: 10.11647/OBP.0316

Cover image: Fall by Tara Shabnavard
Cover design by Anna Gatti

Contents

Acknowledgements	vii
Preface	ix
1. Entering Landscapes of Investigation *Ole Skovsmose*	1
2. Let's Go Shopping *Fanny Aseneth Gutiérrez Rodríguez and Yael Carolina Rodríguez Moreno*	21
3. Media and Racism *Reginaldo Ramos de Britto*	39
4. Bringing the Debate over Marijuana Legalisation into the Mathematics Classroom *Agustín Méndez and Mario Sánchez Aguilar*	57
5. Mathematics Embedded in Community-Based Practices: Landscape of Investigation for Examining Social (In)Justice? *Arindam Bose*	69
6. Aspects of Democracy in Different Contexts of Mathematics Classes *Raquel Milani, Ana Carolina Faustino, Lessandra Marcelly Sousa da Silva, Débora Vieira de Souza-Carneiro, Jeimy Cortés and Reginaldo Ramos de Britto*	95
7. Collaborative Learning within Critical Mathematics Education *Bülent Avcı*	115
8. Global Citizenship *Manuella Carrijo*	133
9. About Unfinishedness, Dreams and Landscapes of Investigation *Daniela Alves Soares*	149

10. A Dialogue in Eternity: Children, Mathematics, and
Landscapes of Investigation   163
Ana Carolina Faustino

11. Inclusive Landscapes of Investigation   185
Ole Skovsmose

12. Meetings amongst Deaf and Hearing Students in the
Mathematics Classroom   197
Amanda Queiroz Moura and Miriam Godoy Penteado

13. Inclusion and Landscape of Investigation: A Case of
Elementary Education   211
Íria Bonfim Gaviolli and Miriam Godoy Penteado

14. Landscapes of Investigation with Seniors   223
*Guilherme Henrique Gomes da Silva, Rejane Siqueira
Julio and Rafaela Nascimento da Silva*

15. The Investigative Approach to Talking about Inclusion in
Mathematics Teacher Education   247
Denner Dias Barros

16. Opening an Exercise: Prospective Mathematics Teachers
Entering into Landscapes of Investigation   257
Raquel Milani

17. The Impact of Income Tax on the Teaching Profession:
A Debate Involving Social Justice   273
*Renato Douglas Gomes Lorenzetto Ribeiro, Daniela Alves Soares,
Adriana de Souza Lima, Lucicleide Bezerra and Edyenis Frango*

18. Critical Mathematics Education in Action: To Be or Not to Be   295
Paula Andrea Grawieski Civiero and Fátima Peres Zago de Oliveira

Contributor Biographies   323
List of Figures   335
List of Tables   337
Index   339

Acknowledgements

The book *Landscapes of Investigation: Contributions to Critical Mathematics Education* was planned, written, and edited during a difficult period of time. The Coronavirus pandemic haunted the world with devastating implications, also for the professional and private lives of the authors of this book. Furthermore, the time was difficult for the two of us, due to Ole's health problems. However, in spite of all difficulties, the book became real. We are grateful for all the support we have received.

We want to thank Pamela Long (in memoriam). She helped us by making a start to the language editing process. We were extremely shocked by the news of her sudden death.

We want to thank the members of the Épura research group for all their help during this long period. We want to thank the authors for their contributions, for their collaborations in the reviewing processes, and for their enthusiasm in creating this book. We want to thank the editors of the book series *Studies in Mathematics Educations and Society* for accepting this book for publication, and thanks to David Kollosche, Brian Lawler, and David Wagner for guiding us through the process. We want to thank Peter Gates for his help and support during the process, and Rosalyn Sword for completing careful copy-editing of the whole manuscript.

August 2022
Miriam Godoy Penteado and Ole Skovsmose

Preface

Landscapes of Investigation: Contributions to Critical Mathematics Education emerged from the collaboration established through the Colloquium of Research in Critical Mathematics Education (Colóquio de Pesquisa em Educação Matemática Crítica). This event took place in 2016 and 2018 at Universidade Estadual Paulista (Unesp) in Rio Claro, and in 2019 at the University of São Paulo (USP) in São Paulo. Further Colloquium events are planned when the pandemic is under control.

In the Colloquium, most of the participants are young researchers dedicated to the further development of critical mathematics education. Many of them come from Brazil, but some also join (sometimes virtually) from Chile, Colombia, India, Mexico, and the USA. Several participants are members of the Épura research group, which is coordinated by the two of us (Miriam Godoy Penteado and Ole Skovsmose). This group was founded in 2008 and is associated with Unesp in Rio Claro.[1] Épura members are mainly Master's and PhD students and researchers working within inclusive education inspired by critical mathematics education. *Landscapes of Investigation* emerged through the shared efforts of the Colloquiums and the Épura Group.

The notion of landscapes of investigation has been developed over more than twenty-five years, and this book draws on this entire process. In a crucial way, the book contributes to the further development of the notion of landscape of investigation, both with respect to its theoretical features and its practical impact; and therefore also to the general evolution of critical mathematics education. The creation of landscapes of investigation is an attempt to organise educational processes in such a way that they allow students and teachers to get involved in

---

1 More detail about Épura in https://igce.rc.unesp.br/#!/departamentos/educacao-matematica/grupos-de-pesquisa/epura/apresentacao/

explorative processes guided by dialogical interactions. It is an attempt to address forms of social injustice by means of mathematics. It is also an attempt to promote a critical conception of mathematics, challenging the assumption that mathematics represents objectivity and neutrality.

One of the initial inspirations for recognising education as a critical and political force came from Paulo Freire, when he in 1968 published *Pedagogia do Oprimido*. At that time Brazil suffered under a military dictatorship, so the following formulations of critical education had to maintain a clandestine format. Later, critical education acquired a diverse range of manifestations, and a huge number of educational practices, as well as books, articles, doctoral dissertations, and Master's theses.

In Brazil one also finds a wealth of important contributions to the further development of critical mathematics education. Many teachers and schools have engaged students in project work in mathematics and addressed controversial social and political issues. Critical mathematics education has also been documented by a huge number of publications.

If we consider the contributions by people associated with the Épura research group, we find studies that present a variety of landscapes of investigation, address dialogic interactions in the classroom, consider students in marginalised positions, and challenge upper-middle-class stereotypes, as well as many other issues important for critical mathematics education. Besides producing dissertations and theses, the Épura research group has engaged in developing educational practices where teachers, university students, and researchers are collaborating. The list of productions by the Épura Group is available on the internet. It shows that fifteen doctoral dissertations and eleven Master's theses have been defended, and also that the Épura group has engaged in a variety of other educational activities.[2]

If we look further around in Brazil, one finds many more contributions to critical mathematics education. As a brief illustration we can refer to three recently published books addressing mathematics teacher education from the perspective of critical mathematics education. In 2021, Ana Karina Baroni, Andrei Hartmann, and Claudia Carvalho published *Uma Abordagem Crítica da Educação Financeira na Formação do Professor da*

---

2   See https://igce.rc.unesp.br/#!/departamentos/educacao-matematica/grupos-de-pesquisa/epura/producoes/

Matemática (Editora Appris); the same year, Guilherme Silva, Iranete Lima, and Fanny Aseneth Gutiérrez Rodríguez published *Educação Matemática Crítica e a (In)justiça Social: Práticas Pedagógicas e Formação de Professores* (Mercado de Letras). In 2022, Paula Civiero, Raquel Milani, Aldinete Silvino Lima, and Adriane Souza Lima published *Educação Matemática Crítica: Múltiplas Possibilidades na Formação de Professores que Ensinam Matemática* (Editora da SBEM). These three edited books contain more than thirty different chapters, providing examples of educational practices, including the formation of landscapes of investigation. The books are also rich in making theoretical connections. Also in 2022, João Luiz Muzinatti published *Matemática: Verdade apaziguadora* (Editora Appris). He presents detailed examples of landscapes of investigation, which he has tried out in his own practice. Many of the authors of these books have joined the Colloquium of Research in Critical Mathematics Education, and some have also been members of the Épura group.

In 2013 Daniela Alves Soares published *O Ensino de Matemática em um Perspectiva Crítica: Dimensões Teóricas e Acadêmicas* (Novas Edições Acadêmicos) where she provided an overview of studies, most of them by Brazilian authors, significant for critical mathematics education. Her list of references includes around 100 titles. If we were to update such an overview today, this number would become much larger.

If we look at other Latin American countries like Columbia, Chile, and Argentina, we also see an impressive increase in contributions to critical mathematics education. Some of these are made by people who joined the Colloquium in Research in Critical Mathematics Education, and several of them also have a chapter in the present book.

The chapters in *Landscapes of Investigation: Contributions to Critical Mathematics Education* add to the further development of critical mathematics education. Chapter 1 presents the context of how the very notion of landscape of investigation was formulated, and how it embodies an attempt to create learning environments that stimulate dialogic interactions and critical activities. Chapters 2–5 present examples of such landscapes developed around economic issues, problems of racism, discussions about marijuana legalisation, and issues of social justice and injustice. Chapters 6–10 expand on the discussion of landscapes of investigation by addressing issues of democracy, collaborative learning, global citizenship, unfinishedness, dreams, and dialogue. The notion of landscape of investigation is not only directed towards the creation of educational learning environments, but also towards deep educational, sociological, and philosophical ideas. Chapters 11–14 pay particular attention to inclusive mathematics education by presenting the concept of inclusive landscapes of investigation and by exploring possibilities for establishing mathematics education as meetings amongst differences. One finds studies of classrooms where deaf and hearing students are working together, and classrooms including students with autism spectrum disorder. Chapters 15–18 concentrate on mathematics teacher education and university studies in mathematics. The chapters present landscapes of investigation that incorporate profound social, political, and mathematical complexities. A way of preparing mathematics teachers for working with landscapes of investigation, together with their students, is for them to work with such landscapes as part of their own education.

The chapters in this book explore new educational possibilities. In this way, they contribute to the further development of critical mathematics education. We hope that the reader will feel as inspired as we did when seeing all these different ideas, reflections, and suggestions brought together.

Miriam Godoy Penteado and Ole Skovsmose

1. Entering Landscapes of Investigation

Ole Skovsmose

> Landscapes of investigation provide different learning environments from the paradigm of exercises. The move from exercises into landscapes of investigation is a move into new risk structures, but is also a move into new possibilities. When entering into landscapes of investigation, one cannot expect the communication to follow predictable patterns. Here, one enters an environment which calls for dialogue. I see dialogue as playing a fundamental role in establishing a critical activity, and as a consequence I consider working with landscapes of investigation to be an important feature of critical mathematics education. The exploration of landscapes is not restricted to particular groups of students. Any group of students can be invited into such landscapes: students at social risk, students in comfortable positions, students with disabilities, senior students, and university students in mathematics. Landscapes of investigation are constructions; they are contested, and they can come to include any kind of controversial issue.

In 1996 when I became a professor at the Royal Danish School of Educational Studies in Denmark, I was invited to give a lecture to a large group of mathematics teachers. As it happened, I did not have much time to prepare my presentation, but I felt confident that it would all come out well. I do not remember the exact title of the lecture, but the topic was project work in mathematics education, and how this might relate to the outlook of critical mathematics education. I had very many examples to draw on with respect to elementary and secondary education, as well as to university education. Since 1982 I had been working at Aalborg University, where all study programmes, including mathematics, were project-based.

However, I could not stop myself from worrying a little. At that time in Denmark, the discussion of project work had taken on a predictable format. It was emphasised again and again that the starting point of such projects had to be embedded in real-life problems—if possible, problems formulated by the students themselves—and that the students had to be working in groups. The teachers had heard such recommendations many times before, and this was exactly what I had planned to say. The teachers also knew in advance what questions they were going to ask, and so did I: how do you address the solutions of quadratic equations through real-life problems? How do you find time to cover the curriculum through project work? What do you do if the students do not like working in groups? How do you prepare the students for tests and exams?

Just before the lecture I began to question what I was planning to present when I met with two colleagues from the School of Educational Studies, Lisser Rye Ejersbo and Michael Wahl Andersen. We talked about what I was going to say, and I made some lines, circles, and arrows on the blackboard in their office. This turned out to be the first wobbling draft of the diagram shown in Table 1.

I do not remember how the exact expression "landscape of investigation" came to emerge, but I remember clearly the atmosphere that developed during the lecture. Nobody knew what a landscape of investigation was, nor did I. Anyway, I presented examples, and questions and suggestions came from the teachers. We all became very enthusiastic. We participated in an intensive process of pedagogical imagination, and the notion of landscape of investigation was born in that cauldron of collaboration. The notion provided a way of talking about educational possibilities, about project work, and about inquiry processes, without being assertive about educational principles and priorities.

In the following, I will refer to the *diagram* that I have used many times in my presentations of landscapes of investigation. I will present *dialogue* as being crucial when working with landscapes, and I will address the notion of *critique* in order to capture the way landscapes of investigation relate to critical mathematics education. I will address *different groups of students* that can come to explore landscapes of investigation. Finally, I will highlight that landscapes are *contested constructions*.

1. The Diagram

Soon after making the presentation at the Royal Danish School of Educational Studies, I was invited to give another lecture in Norway, and at this lecture I presented the diagram—this time well-prepared, with nice straight lines showing different learning environments. In the diagram I made a distinction between "paradigm of exercises" and "landscapes of investigation". Furthermore, I distinguished between three different forms of references one can operate with within mathematics education: namely, to mathematics itself, to a semi-reality, and to real-life situations. With these distinctions one can identity six different learning environments that I will also refer to as learning *milieus*. The diagram helped me to highlight several points (see Fig. 1).

First, landscapes of investigation provide quite different learning environments to the paradigm of exercises, which I also refer to as the school mathematics tradition.[1] This tradition is characterised by the students working with preformulated exercises, each of which has one, and only one, correct solution. By talking about landscapes of investigation, one prepares for leaving aside the exercise paradigm. One can oblige students to solve exercises, but one cannot force them to undertake investigations. Therefore I have always emphasised that students need to be *invited* into a landscape of investigation. Such a learning environment calls for a different pattern of student-teacher interaction. The teacher does not need to "teach", but can serve as supervisor, and dialogue can structure the communication.

---

1   For a characterisation of this tradition, see Skovsmose and Penteado (2016).

Table 1. Learning milieus.

	Paradigm of exercises	Landscapes of investigation
References to mathematics	(1)	(2)
References to a semi-reality	(3)	(4)
Real-life references	(5)	(6)

Second, leaving the exercise paradigm can be done in a number of different ways and in different steps. Often, the move has been presented as a huge jump from doing exercises straight to doing project work embedded in real-life problems. We are dealing here with a jump from the milieus (1) or (3) to the milieu (6). Considering the reality of schools, this jump confronts several obstructions. For many teachers, it might be an overwhelming experience. However, moving from the exercise paradigm to landscapes of investigation can also be undertaken with small steps. As an illustration of such steps, I sometimes refer to the function:

$$f(x) = ax^2 + bx + c$$

Many exercises have been formulated with reference to this function. Given specific values for a, b, and c, students are asked to draw the graph of the function, to calculate the value of x that results in $f(x)$ having the value of 4, and to find where it intersects a given line. From the textbooks, the students will come to know about the significance of the parameters a and c. But what about the parameter b? How does the value of b influence the position of the graph for f? By raising this question, one is entering into a small landscape of investigation. One can imagine that the students start undertaking some experimentation: fixing the value of a and c, and changing the value of b to see what might happen. How then are we to formulate what can be observed? In fact, any classic mathematical exercise can provide similar openings towards broader landscapes of investigation.[2]

---

[2] Such processes have been described by Milani in Chapter 16, "Opening an Exercise". Wherever I refer to a chapter, it is always a chapter in this book.

Third, the diagram refers to mathematics *per se*, semi-reality, and real-life situations. Broadening the space of possible references helps us to identify different teaching-learning possibilities. Reference can be made to pure mathematical problems. Such references can be in terms of exercises (Milieu 1), but also as investigative tasks (Milieu 2). Traditional exercises (Milieu 3) may refer to families going on holiday, to somebody buying a car, to the size of a parcel, etc. But we are not dealing with any real family, car, or parcel. Everything is made up by the textbook author. Here we only deal with invented realities, which I refer to as semi-realities. However, semi-realities can also provide contexts for landscapes of investigation (Milieu 4). Milieu (5) is composed of exercises using real numbers referring to, for instance, degrees of unemployment, development of house prices, or social exclusion due to racism or sexism.[3] The same kind of references characterises Milieu (6); here, however, they are organised in terms of proper investigations. The point of the diagram is to indicate the existence of a variety of learning environments, and in this way to support a pedagogical imagination.[4]

Fourth, one can move among the learning milieus in different ways. When I give presentations to teachers, I have sometimes asked which environment they were working in, say, last Monday. The diagram can function as a stimulus for discussing educational possibilities and difficulties. Where did the teachers experience problems? Where did they feel most comfortable? Where did the students become most involved? One can use the diagram as a starting point for reflections on educational practices. It can also be used as support for educational planning. It can help the teacher in identifying what could be an appropriate route to take with the class among the different learning environments. Would it be a good idea to start in (5) before moving on to (6)? And after being engaged in (6), would it be a good idea to move to (1) for a while?

Fifth, several times it has been highlighted that risks and possibilities are different with respect to the different learning environments.[5] For teachers to feel well prepared, it is easier to operate in milieus (1) or (3) than in any landscape of investigation. Milieu (5) can cause some

---

3   In Frankenstein (1989) one can find many of such exercises.
4   For a discussion of pedagogical imagination, see Skovsmose (2011b).
5   The notion of risk with respect to mathematics education has been explored in great detail by Penteado (2001) and further elaborated by Biotto Filho (2008).

uncertainty with respect to what information to use; however, with respect to milieus (1) and (3), the teacher can check out solutions in advance. Methods can be clarified, and solutions can be verified. As soon as one enters landscapes of investigation, uncertainties will emerge and predictions of what will happen become impossible. The teacher can still be prepared, but not in the same way. The students' questions cannot be predicted, and adequate answers cannot be formulated in advance. The move from the exercise paradigm into landscapes of investigation is a move into new risk structures. It is, as well, a move into new educational possibilities.

After my first presentations of landscapes of investigations, there followed many more lectures and publications, but to a large extent these were repetitions.[6] However, many colleagues have contributed significantly to the development of the concept of landscapes of investigation.[7]

2. Dialogue

Dialogue is a captivating notion, as it automatically seems to bring about positive connotations. Who would argue against establishing more dialogue in the classroom? However, dialogue is an open concept that cannot be associated with a single and clear interpretation. So I had better say a bit more about how I intend to use—and not to use—the notion, and I want to highlight the following three points.

First, I refer to dialogue as it may emerge in educational processes. I relate dialogue to processes of coming to know something and to learning inquiries. Thus there are many uses of the notion of dialogue that I do not consider. In his hard-boiled detective novels, Dashiell Hammett wrote a range of captivating dialogues to which, when the

---

6   The text about landscapes of investigations was published in proceedings and pre-prints in Danish (Skovsmose, 1998, 1999) and in English (Skovsmose, 2000a). The text was published in Portuguese (Skovsmose, 2000b, 2001b), and in Spanish (Skovsmose, 2000c, 2012). The English version was published in a journal (Skovsmose, 2001a), as Chapter 1 in Skovsmose (2014), and in a shortened version as a chapter in a book (Skovsmose, 2002). It also became a regular chapter in a book in Danish (Skovsmose, 2003).
7   See, for instance, Biotto Filho (2015); Biotto et al (2017); Britto et al. (2017); Milani et al. (2017); Moura (2020); Oliveira et al. (2017); Roncato (2015); Silva et al. (2017); and Voltolini and Kaiber (2017).

books were turned into films, Humphrey Bogart added an extra dose of tough chauvinism. However, I am not going to refer to dialogue in terms of conversations as they may be presented in books, plays, or films. *Second*, I see dialogue as an open process, meaning that one cannot expect a dialogue to follow any specific pattern. In a dialogue one reacts to what has just been said, implying that the horizon of a dialogue is always shifting. The course of a dialogue is unpredictable. *Third*, I see dialogue as incorporating features of equity. In many contexts—the military, companies, organisations—the position of the people involved influences the patterns of conversation. In a dialogue, however, it is the content of what is being said, rather than the position of the speaker(s), that plays the principal role. This observation brings me to highlight equity as being important for characterising a dialogue.[8]

The exercise paradigm does not create much space for dialogue, whereas in contrast landscapes of investigation do. An exercise defines a task for the students to complete, and this structures the conversation between students and teachers, as well as among students. Within the exercise paradigm, it is not considered relevant for the students to change the formulation of an exercise and start trying to solve a simplified version of the problem. This highly relevant mathematical strategy is blocked by the school mathematics tradition. An exercise is not an invitation for undertaking an investigation; rather, it operates as an order. For solving an exercise, one needs to proceed along a narrow predefined one-way route, and the conversation that takes place along such a route is itself predefined. The students can ask whether a certain procedure is the proper one, and if a certain answer is correct. The teacher can confirm, or correct, or clarify.

When working with landscapes of investigation, one cannot expect the communication to follow particular predetermined and well-planned routes. The conversation will be open-ended and dialogical, as inquiry processes are open-ended. I find that important qualities of

---

8   Dialogue was related to processes of investigation, unpredictability, and equity in Alrø and Skovsmose (2004). In the present text, however, I have made some terminological modifications. For profound discussions of dialogue, see Milani (2015) and Faustino (2018). The notion of dialogue has also been addressed by Faustino in Chapter 10, "Dialogue in Eternity", while the notion of collaboration has been addressed by Avcı in Chapter 7, "Collaborative Learning within Critical Mathematics Education".

learning emerge through dialogic processes. This observation brings me to highlight the importance of characterising *dialogic acts* as well as *non-dialogic acts*. The identification of such acts is important for referring to processes of communication as being dialogical or not.

Helle Alrø and I (2004) have characterised eight *dialogue acts*. By *getting in contact*, we refer to the process of establishing emotional contact and to showing interest in each other. *Locating* refers to attempts to grasp the overall concern of the other. It refers to a clarification of topics and concerns of the conversation. *Identifying* is a process through which one tries to be more specific about the issues one is addressing. With reference to mathematics, one can, for instance, identify which equation one is going to investigate. *Advocating* means providing arguments for a given case. This could refer to the process of proving a mathematical statement. *Thinking aloud* occurs when one shows others how one is reasoning. Here we are dealing with a process of making thoughts not only audible, but also visible in the form of figures and sketches. *Reformulating* serves an important role in coming to understand each other better. When one has been listening to an explanation, one may try to repeat it, although doing so in one's "own words". *Challenging* means questioning a certain statement or perspective, but it does not mean attacking the other person. *Evaluating* is an important dialogic act, as it is necessary to reflect on the steps taken in a dialogue.[9]

Ana Carolina Faustino and I (2020) have characterised eight *non-dialogic* acts. *Ignoring* means disregarding something being stated, and just continuing the conversation as if nothing has happened. *Distorting* can be established through a reformulation, which provides a caricature of what has been stated. *Confronting* can take the form of a direct negation of a statement, without considering the arguments that have led to the statement. *Ridiculing* can be brutal. It addresses the person participating in the conversation, rather than its content. *Disqualifying* is also a way of attacking a person, but it could take a different format than ridiculing. Disqualifying means, for instance, pointing out that the person does not have sufficient knowledge for addressing the issue in question. *Excluding* is different from ignoring. While ignoring means disregarding a person, exclusion is of a more profound nature. It takes place not just during moments, but during periods of time. *Stigmatising*

---

9   See also the related discussion of inquiry gestures in Milani and Skovsmose (2014).

can take place with reference to gender, race, religion, sexual orientation, special educational needs, home language, and social class; it can serve as a justification for an exclusion. *Lecturing* refers to a situation where a person dominates a conversation to elaborate a definite point of view without considering its actual relevance for the present discussion.

Taken together, the eight dialogic and eight non-dialogic acts help in analysing communicative processes. Such a process can be called dialogical in the case where it includes a density of dialogic acts. If a high frequency of non-dialogic acts occurs, the conversation must be characterised otherwise. A fluctuation between dialogic and non-dialogic acts can often be observed, not least in the mathematics classroom. While different forms of communication, also of non-dialogical nature, can operate within the exercise paradigm, dialogical processes must be established in landscapes of investigation. I see *investigative processes as dialogic processes*. When one enters a landscape of investigation, one needs to consider how to let dialogic acts achieve priority over non-dialogic acts.

3. Critique

Why should one be interested in dialogue? As discussed above, I find that important qualities of learning emerge through dialogic processes. To be more specific, I find critical activities to be routed in dialogic processes.

The notion of critique has a deep and complex history of which I want to highlight only two moments, namely the publication of Immanuel Kant's (1933) work *Critique of Pure Reason* in 1781, and the publication of Karl Marx's (1992, 1993a, 1993b) work *Capital: A Critique of Political Economy*, the first volume of which appeared in 1867. Kant shared the outlook of Enlightenment, according to which knowledge is the true source of human progress. But what is knowledge? Kant addressed this question by trying to clarify the nature, as well as the limits of, human knowledge. In this way, he formulated a critique of knowledge as an epistemic enterprise. Marx provided a radical addition to this conception of critique. He wanted not only to criticise economic and political theories, but also the very economic and political structures themselves. To Marx, critique was not only an epistemic, but also a socio-political activity.

When I talk about critique, I have both the epistemic and the socio-political aspect in mind. Thinking of Kant and Marx, these two aspects appear deeply separated, but this need not be the case. Sometimes Michel Foucault has been referred to as the new Kant. One reason for this comparison is that Foucault also addresses our basic conceptual structures, through which we formulate what we assume to know. He wanted to show how such categories are formed as part of complex historical processes, and that they may incorporate all kinds of presumptions and preconceptions. What Kant thought of as pure epistemic categories, Foucault revealed as conglomerates of preconceptions. In this way, Foucault formulated a radical epistemic critique, which simultaneously was a powerful socio-political critique.[10]

I see dialogue as *playing a fundamental role in establishing a critical activity*. In making this claim, I cannot refer to Kant nor to Marx, as neither of them paid attention to dialogic processes. The relationship between critique and dialogue was articulated long after the publication of their works. Nor did Foucault pay much attention to the notion of dialogue. With reference to connections between dialogue and the epistemic aspect of critique, I can refer to the work of Imre Lakatos, and with reference to the connections between dialogue and the socio-political aspect of critique, I can refer to the work of Paulo Freire.

In the first pages of *Proofs and Refutations*, Lakatos (1976) presents a proof of Euler's Polyhedron Theorem. It is common to assume that a mathematical proof provides the concluding part of a mathematical exposition. However, Lakatos shows that a proof is just a first step in an ongoing inquiring process. Proofs lead to refutations, which in turn lead to formulations of new mathematical hypotheses, as well as to new mathematical concepts. The whole process is presented by Lakatos as a dialogue taking place within a fictive mathematics classroom, which represents a mathematical research community. To me, Lakatos shows how an inquiring process in mathematics can take a dialogical format, and also that dialogue plays an important role for formulating an epistemic critique in mathematics. Through the dialogue, assumed mathematical truths are questioned, and mathematical assumptions are modified, if not falsified.

---

10   See, for instance, Foucault (1989, 1994).

As a point of departure for a "pedagogy of the oppressed", Paulo Freire (1972) wants to identify and investigate cases of social injustice, and in this way, to prepare for liberating actions. To Freire, dialogue plays a crucial role in such an education. He highlights the importance of taking as a starting point "generative themes", which represent situations that are well known to the students. One can think of generative themes as examples of landscapes of investigation of the type (6). Through dialogue between students and teachers, the generative theme is explored, and the nature of patterns of oppression and exploitation might be identified. This prepares them for political actions. According to Freire, it is essential that the exploration of generative themes achieves a dialogical format, as he sees an intimate relationship between dialogue and the formations of a lived-through departure for political actions.

I find any critique of either an epistemic or a socio-political nature to be rooted in dialogical processes. Due to the connection between dialogue and critique, I find it important that critical mathematics education establishes educational processes in a dialogic format. To operate in landscapes of investigation is a way of doing so. Thus my interest in landscapes of investigations is embedded in my concern for establishing learning milieus that invite dialogic interactions and, in this way, stimulate critical activities.[11]

4. Different Groups of Students

The exploration of landscapes of investigation is not restricted to certain groups of students. Any group of students can be invited into such landscapes, and from the perspective of critical mathematics education, it is important to do so. I will highlight this point by being explicit about

---

11 I use the notion of critical mathematics broadly to also include mathematics education for social justice. Very many people have contributed to the further development of critical mathematics education. For recent contributions, I can refer to Alrø et al. (2010), Andersson and Barwell (2021), Avci (2018), Bartell (2018), Ernest et al. (2015), Frankenstein (2012), Greer et al. (2009), Gutstein (2016, 2018), Skovsmose (2011a, 2014), and Skovsmose and Greer (2012). For a broader discussion of the notion of citizenship, see Chapter 8, "Global Citizenship" by Carrijo.

students at social risk, students in comfortable positions, students with disabilities, senior students, and university students in mathematics.[12]

Students at social risk. Eric Gutstein works with students at social risk, with whom he addresses topics like election, immigration, deportation, foreclosure, the spread of HIV-AIDS, criminalisation of young people of colour, racism, and sexism (see Gutstein, 2016, 2018). In each case mathematical investigations can help to reveal features of oppression, exploitation, and injustice. Gutstein is deeply inspired by Freire, and he sees his education as one possible interpretation of a "pedagogy of the oppressed". He forms alliances with students who suffer from economic oppression and racist discourses, and he tries to turn mathematics education into a source for liberating actions. Many other contributions to critical mathematics education share this concern with respect to students at social risk (see, for instance, Frankenstein, 2012). I find it extremely important to open up landscapes of investigation for these groups of students in order to act against all forms of oppression.[13]

Students in comfortable positions. One can also consider a very different group of students, which I refer to as students in "comfortable positions". I have in mind students who belong to the well-protected layers of society. For such groups of students, it is also important to provide landscapes of investigation, through which one can address social and economic injustices. João Luiz Muzinatti (2022) tries to challenge general assumptions that make up a traditional middle-class outlook. As an example of such an approach, he addresses *Bolsa Familia*, which is a Brazilian system of family support. *Bolsa Familia* has faced much middle-class critique, such as: we are paying an awful lot of money to people that, as a consequence, do not want to work. By means of mathematics, Muzinatti challenges the content of such general claims. Working for social justice not only concerns groups of people suffering injustices, it also concerns groups who might benefit from injustices done to others. As this example illustrates, an important strategy of such

---

12 In the following, I draw on Skovsmose (2016).
13 Racism has been addressed by Britto in Chapter 3, "Median and Racism". Issues related to students at risk have been addressed by Bose in Chapter 5, "Mathematics Embedded in Community-Based Practices", and by Soares in Chapter 9, "About Unfinishedness, Dreams and Landscapes of Investigation".

an approach is to establish dialogic relationships, as Muzinatti did with his own students.[14]

Students with disabilities.[15] Lessandra Marcelly (2015) addresses the learning of mathematics by blind students. The obvious question here is: what could be a landscape of investigation for blind students? However, Marcelly moves beyond this question by asking: how can one provide learning environments where blind and sighted students can work together? By addressing this question, one acknowledges a principal concern of inclusive education. Amanda Moura (2020) also explores such education by researching situations where deaf and hearing students are working together. The idea of inclusive education highlights the importance of establishing meetings across differences. In this respect, one can consider any kind of differences with respect to seeing, hearing, or abilities.[16] The differences could also be with respect to political opinions, economic resources, and cultural belongings. Inclusive education is a preoccupation of critical mathematics education, as inclusion might establish new possibilities for dialogical encounters and critical activities.[17]

Senior students. Luciano Feliciano de Lima (2015) worked with senior students, who could be retired bank assistants, shopkeepers, or street workers. Lima's project was not part of any formal educational programme, and the students joined out of personal interest. They became engaged in different landscapes of investigation. One concerned geometric figures, and notions like symmetry, congruence, and reflection were explored through experimentations with mirrors. More complex mathematical properties were addressed as well; for instance, Euler's Polyhedron Theorem. The daily newspapers were also studied,

---

14 For working with students in comfortable positions, see Chapter 4, "Bringing the Debate over Marijuana Legalisation into the Mathematical Classroom" by Méndez and Aguilar.

15 The very notion of "disability" is problematic. This has been highlighted by Marcone (2015), who coined the notion of "deficiencialism". This refers to a web of preconceptions concerning what a certain group of people can and cannot do.

16 For further discussions of inclusive education, see Chapter 12, "Meetings Amongst Deaf Students in the Mathematics Classroom" by Moura and Penteado; Chapter 13, "Inclusion and Landscape of Investigation" by Gaviolli and Penteado; and Chapter 15, "The investigative Approach to Talking about Inclusion in Mathematics Teacher Education" by Barros.

17 For a further discussion of this point, see my Chapter 11, "Inclusive Landscapes of Investigation". See also Skovsmose (2019).

since they are loaded with numbers and figures, and the senior students learned to interpret such information. In this way they gained access to a range of economic and political issues that they had previously tended to ignore. This is an explicit example of getting students to read the world through mathematics, which is a principal concern of critical mathematics education.[18]

University students in mathematics. For centuries mathematics has been celebrated as the language of science, ensuring the most reliable way of capturing and presenting scientific observations and insight. To a large extent this celebration defines the format of university studies in mathematics. Typically, these studies are entirely focussed on content-matter issues and organised according to the exercise paradigm. From the perspective of critical mathematics education, it is crucial that university studies in mathematics create space for reflections on mathematics, and in particular on what putting mathematics into action might imply. Landscapes of investigation could create an opening for both epistemic and socio-political forms of critique, but they are rarely developed with university studies in mind. However, one can find inspiration for constructing such landscapes in the literature about problem-based learning in mathematics, as discussed by Débora Vieira de Souza-Carneiro (2021), and in project work in mathematics as analysed by Renuka Vithal, Iben Maj Christiansen, and Ole Skovsmose (1995).[19]

5. Contested Constructions

Table 1 provides a preliminary classification of learning milieus: three related to the exercise paradigm and three related to landscapes of investigation. Naturally, we are not dealing with any genuine classification, and one could imagine many different kinds of intermediate or overlapping milieus. One might also think of other

---

18 See also Chapter 14, "Landscapes of Investigation with Seniors" by Silva, Julio, and Silva.
19 For establishing landscapes of investigation for university students, see Chapter 18, "Critical Mathematical Education in Action" by Civiero and Oliveira. For a discussion of mathematics teacher education, see Chapter 17, "The Impact of Income Tax on the Teaching Profession" by Ribeiro, Soares, Lima, Bezzerra, and Frango. See also the discussion of banality of mathematical expertise in Skovsmose (2020).

ways of characterising landscapes of investigation, and let me just refer to three such ways: inclusive, contentious, and exemplary landscapes.

Inclusive landscapes of investigation. As already pointed out, a general concern of inclusive education, also shared by critical mathematics education, is to provide learning environments where all students, independent of differences, can learn together. This concern brings us to the idea of constructing inclusive landscapes of investigation: ones which are accessible for everybody. Differences among students should not create specific conditions for entering into and moving around in such landscapes. In inclusive landscapes, the very conception of students as being normal or not normal, or having abilities or a disability, lose significance. It is important to establish relationships across all kinds of differences. One need not only consider differences rooted in terms of abilities, but differences in general, and see them as a universal feature of human life. One can, for instance, consider economic, religious, cultural, or political differences, and create landscapes across such differences.

Contentious landscapes of investigation. The notion of generative themes plays a crucial role in Freire's formulation of a pedagogy of the oppressed. A theme could be "water", and with reference to a particular neighbourhood one can raise questions like: who has access to water? Who has access to clean water? What health problems could be related to the use of polluted water? A generative theme such as "water" can lead to the articulation of controversial socio-political issues. The topic of "the geometry of one's room" could concentrate on making measurements and doing scale-drawings. The same topic, however, could turn into a discussion of conditions for living and of the profound economic inequalities that exist in society. Just as with "water", so too could "the geometry of one's room" become contentious. I see it as important to invite students into contentious landscapes of investigation, as they might frame dialogues about crucial issues and about social injustices.[20]

Exemplary landscapes of investigation. When an educational process is established through a problem and not through a predefined

---

20 As examples of contentious landscapes of investigation, see Chapter 2, "Let's Go Shopping" by Rodríguez and Moreno, Chapter 6, "Aspects of Democracy in Different Contexts of Mathematics Classes" by Milani, Faustino, Silva, Carneiro, Suaréz, and Britto; and Chapter 4, "Bringing the Debate over Marijuana Legalisation into the Mathematical Classroom" by Méndez and Aguilar.

curriculum, one can consider to what extent the problem is exemplary. The techniques one is applying to solve the problem could be exemplary in the sense that they can be used for addressing a range of other problems. However, exemplarity not only concerns methodological, but also content-matter issues. When one addresses water supply in a neighbourhood, one at the same time gets an insight into more general conditions for living. Studying a specific problem can simultaneously provide insight into much more profound socio-political issues. Landscapes of investigation can be exemplary in this way. As an example, a landscape of investigation related to Euler's Polyhedron Theorem can be exemplary with respect to mathematical thinking, to the extent that it stimulates patterns of mathematical arguing, proving, and questioning. The landscape of *Bolsa Familia* can be exemplary with respect to ways of addressing social issues; thus it concerns not only a particular form of family support, but also general issues about social welfare and justice.

Landscapes of investigation—whether they are inclusive, contentious, or exemplary—are constructions. Teachers have played crucial roles in the construction of the very many landscapes in which I have participated. One step in the construction is to recognise a possible terrain for investigations; an important further step is to make available relevant information, statistics, facts, and figures that one can invite students to explore. In this process, a range of uncertainties will arise: will the students find it interesting to investigate the landscape? Or should the students themselves decide which landscape to explore? Should the landscape be contentious or not, inclusive or not, exemplary or not? What socio-political issues could be addressed when working with the landscape? What issues of social injustice are placed on the agenda? The construction of any landscape of investigation can be contested.

Acknowledgements

Several of the ideas and some of the formulations that I have used are also found in Skovsmose (2016, 2019, in print). I want to thank Denner Barros, Manuella Carrijo, Ana Carolina Faustino, Amanda Queiroz Moura, João Luiz Muzinatti, Miriam Godoy Penteado, Célia Regina Roncato, Daniela Alves Soares, and Débora Vieira de Souza for their helpful comments and suggestions.

References

Alrø, H., Ravn, O. & Valero, P. (Eds) (2010). *Critical mathematics education: Past, present, and future.* Sense Publishers.

Alrø, H. & Skovsmose, O. (2004). *Dialogue and learning in mathematics education: Intention, reflection, critique.* Kluwer Academic Publishers.

Andersson, A. & Barwell, R. (Eds) (2021). *Applying critical mathematics education.* Brill and Sense Publishers.

Avci, B. (2018). *Critical mathematics education: Can democratic mathematics education survive under neoliberal regime?* Brill and Sense Publishers.

Bartell, T. G. (Ed.) (2018). *Towards equity and social justice in mathematics education.* Springer.

Biotto Filho, D. (2008). *O desenvolvimento da matemacia no trabalho com projetos.* Master's thesis. Universidade Estadual Paulista (Unesp).

Biotto Filho, D. (2015). *Quem não sonhou em ser um jogador de futebol? Trabalho com projetos para reelaborar foregrounds.* Doctoral dissertation. Universidade Estadual Paulista (Unesp).

Biotto Filho, D., Faustino, A. C. & Moura, A. Q. (2017). Cenários para investigação, imaginação e ação. *Revista Paranaense de Educação Matemática, 6*(12), 64–80.

Britto, R. R., Moura, A. Q., Nascimento, C. A. F., Roncato, C., Biotto Filho, D. & Figueiredo, M. O. R. (2017). Cenários para investigações nas salas de aulas de matemática de escolas brasileiras. *Revista Paranaense de Educação Matemática, 6*(12), 371–396.

Ernest, P., Sriraman, B. & Ernest, N. (Eds) (2015). *Critical mathematics education: Theory, praxis, and reality.* Information Age Publishing.

Faustino, A. C. (2018). *Como você chegou a esse resultado?: O processo de dialogar nas aulas de matemática dos anos iniciais do Ensino Fundamental.* Doctoral dissertation. Universidade Estadual Paulista (Unesp).

Faustino, A. C. & Skovsmose, O. (2020). Dialogic and non-dialogic acts in learning mathematics. *For the Learning of Mathematics, 40*(1), 9–14.

Foucault, M. (1989). *The archaeology of knowledge.* Routledge.

Foucault, M. (1994). *The order of things: An archaeology of the human sciences.* Vintage Books.

Frankenstein, M. (1989). *Relearning mathematics: A different third R—radical maths.* Free Association Books.

Frankenstein, M. (2012). Beyond math content and process: Proposals for underlying aspects of social justice education. In A. A. Wager & D. W. Stinson (Eds), *Teaching mathematics for social justice: Conversations with mathematics educators* (pp. 49–62). NCTM, National Council of Mathematics Teachers.

Freire, P. (1972). *Pedagogy of the oppressed*. Penguin Books.

Greer, B., Mukhopadhyay, S., Powell, A. B. & Nelson-Barber, S. (Eds) (2009). *Culturally responsive mathematics education*. Routledge.

Gutstein, E. (2016). "Our issue, our people—Math as our weapon": Critical mathematics in a Chicago neighborhood high school. *Journal for Research in Mathematics Education, 47*(5), 454–504.

Gutstein, E. (2018). The struggle is pedagogical: Learning to teach critical mathematics. In P. Ernest (Ed.), *The philosophy of mathematics education today* (pp. 131–143). Springer.

Kant, I. (1933). *Critique of pure reason* (N. Kemp Smith, Trans.). MacMillan. (Original work in German published in 1781)

Lakatos, I. (1976). *Proofs and refutations*. Cambridge University Press.

Lima, L. F. (2015). *Conversas sobre matemática com pessoas idosas viabilizadas por uma ação de extensão universitária*. Doctoral dissertation. Universidade Estadual Paulista (Unesp).

Marcelly, L. (2015). *Do improviso às possibilidades de ensino: Um estudo de caso de uma professora de matemática com estudantes cegos*. Doctoral dissertation. Universidade Estadual Paulista (Unesp).

Marcone, R. (2015). *Deficiencialismo: A invenção da deficiência pela normalidade*. Doctoral dissertation. Universidade Estadual Paulista (Unesp).

Marx, K. (1992, 1993a, 1993b) *Capital: A critique of political economy, I–III*, Penguin Classics.

Milani, R. (2015). *O processo de aprender a dialogar por futuros professores de matemática com seus alunos no estágio supervisionado*. Doctoral dissertation. Universidade Estadual Paulista (Unesp).

Milani, R. & Skovsmose, O. (2014). Inquiry gestures. In O. Skovsmose, *Critique as uncertainty* (pp. 45–56). Information Age Publishing.

Milani, R., Civiero, P. A. G., Soares, D. A. & Lima, A. S. (2017). O diálogo nos ambientes de aprendizagem nas aulas de matemática. *Revista Paranaense de Educação Matemática, 6*(12), 221–245.

Moura, A. Q. (2020). *Cenários para investigação e escola inclusiva: Possibilidades de diálogo entre surdos e ouvintes em aulas de matemática*. Doctoral dissertation. Universidade Estadual Paulista (Unesp).

Muzinatti, J. L. (2022). *Matemática: Verdade apaziguadora (Mathematics: Soothing Truth)*. Appris.Muzinatti, J. L. (2018). *A "verdade" apaziguadora na educação matemática: Como a argumentação de estudantes de classe média pode revelar sua visão acerca da injustiça social*. Doctoral dissertation. Universidade Estadual Paulista (Unesp).

Oliveira, A. A., Santos. L. T. B. & Pessoa, C. A. S. (2017). Do exercício aos cenários para investigação: A aplicação de atividades de educação financeira por professores dos anos iniciais do ensino fundamental em uma escola de Recife, PE. *Revista Paranaense de Educação Matemática*, 6(12), 158–186.

Penteado, M. G. (2001). Computer-based learning environments: Risks and uncertainties for teachers. *Ways of Knowing* 1(2), 23–35.

Roncato, C. R. (2015). *Cenários investigativos de aprendizagem matemática: Atividades para a autonomia de um aprendiz com múltipla deficiência sensorial.* Master's thesis. Universidade Anhanguera de São Paulo.

Silva, G. H. G., Marcone, R., Brião, G. F. & Kistemann Jr., M. A. (2017). Educação matemática crítica e preocupações urgentes: Cenários promovedores de equidade e justiça social. *Revista Paranaense de Educação Matemática*, 6(12), 130–157.

Skovsmose, O. (1998). Undersøgelseslandskaber. In T. Dalvang & V. Rohde (Eds), *Matematik for Alle* (pp. 24–37). Landslaget for matematikk i skolen (LAMIS).

Skovsmose, O. (1999). *Undersøgelseslandskaber* (Publication no. 5). Centre for Research in Learning Mathematics, Royal Danish School of Educational Studies, Roskilde University Centre, Aalborg University.

Skovsmose, O. (2000a). *Landscapes of investigation* (Publication no. 20). Centre for Research in Learning Mathematics, Royal Danish School of Educational Studies, Roskilde University Centre, Aalborg University.

Skovsmose, O. (2000b). Cenários para investigação. *Bolema*, 14, 66–91.

Skovsmose, O. (2000c). Escenarios de investigación. *Revista EMA*, 6(1), 1–25.

Skovsmose, O. (2001a). Landscapes of investigation. *Zentralblatt für Didaktik der Mathematik*, 33(4), 123–132. Reprinted as Chapter 1 in Skovsmose, O. (2014), *Critique as uncertainty* (pp. 3–20). Information Age Publishing.

Skovsmose, O. (2001b). Cenários para investigação. In D. Moreira, C. Lopes, I. Oliveira, J. M. Matos & L. Vicente (Eds), *Matemática e comunidades: A diversidade social no ensino-aprendizagen da matemática* (pp. 26–40). Instituto de Inovação Educacional.

Skovsmose, O. (2002). Landscapes of investigation. In L. Haggarty (Ed.), *Teaching mathematics in secondary schools: A reader* (pp. 115–128). Routledge Falmer.

Skovsmose, O. (2003). Undersøgelseslandskaber. In O. Skovsmose & M. Blomhøj (Eds), *Kan der virkelig passe?* (pp. 143–157). LR-Uddannelse.

Skovsmose, O. (2011a). *An invitation to critical mathematics education.* Sense Publishers.

Skovsmose, O. (2011b). Critique, generativity, and imagination. *For the Learning of Mathematics*, 31(3), 19–23.

Skovsmose, O. (2012). Escenarios de investigación. In P. Valero & O. Skovsmose (Eds), *Educación matemática crítica: Una visión sociopolítica del aprendizaje y la enseñanza de las matemáticas* (pp. 109–130). Centro de Investigación y Formación en Educación, Universidad de los Andes.

Skovsmose, O. (2014). *Critique as uncertainty*. Information Age Publishing.

Skovsmose, O. (2016). What could critical mathematics education mean for different groups of students? *For the Learning of Mathematics, 36*(1), 2–7.

Skovsmose, O. (2019). Inclusions, meetings and landscapes. In D. Kollosche, R. Marcone, M. Knigge, M. G. Penteado & O. Skovsmose (Eds), *Inclusive mathematics education: State-of-the-art research from Brazil and Germany* (pp. 71–84). Springer.

Skovsmose, O. (2020). Banality of mathematical expertise. *ZDM Mathematics Education, 52*(6), 1187–1197. https://doi.org/10.1007/s11858-020-01168-4

Skovsmose, O. (in print). Dialogue, landscapes and critique. In N. Kennedy & E. Marchall (Eds), *Dialogical inquiry in mathematics teaching and learning: A philosophical approach*. LIT Publishers.

Skovsmose, O. & Greer, B. (eds) (2012). *Opening the cage: Critique and politics of mathematics education*. Sense Publishers.

Skovsmose, O. & Penteado, M. G. (2016). Mathematics education and democracy: An open landscape of tensions, uncertainties, and challenges. In L. D. English & D. Kirshner (Eds), *Handbook of international research in mathematics education* (3rd ed., pp. 359–373). Routledge.

Souza-Carneiro, D. V. (2021). *A matemática em ação e o ensino de cálculo: Possibilidades no ensino superior por meio da aprendizagem baseada em problemas*. Doctoral dissertation. Universidade Estadual Paulista (Unesp).

Vithal, R., Christiansen, I. M. & Skovsmose, O. (1995). Project work in university mathematics education: A Danish experience: Aalborg University. *Educational Studies in Mathematics 29*(2), 199–223.

Voltolini, L. & Kaiber, C. T. (2017). Cenários para investigação: Ambientes de aprendizagem matemática na educação escolar indígena. *Revista Paranaense de Educação Matemática, 6*(12), 187–202.

2. Let's Go Shopping

Fanny Aseneth Gutiérrez Rodríguez and Yael Carolina Rodríguez Moreno

---

"Let's Go Shopping" is a landscape of investigation about economic and financial education framed by the socio-critical perspective of mathematical modelling. The landscape developed as a result of discussions among students from a school in Bogotá about high-interest charges and loans. Our aim was to identify possibilities and challenges in the use of a socio-critical perspective on mathematical modelling when promoting citizenship education. To do this, we addressed the discussions by the students and considered students' intentions, backgrounds, and foregrounds. The landscape included investigations of families' economic conditions, as the students related the mathematics classroom activities to their families' financial issues.

The school day is underway in the Bogotá capital city of Colombia in a school located in Neighbourhood 18 called Rafael Uribe Uribe, and the ninth-grade students are preparing to receive their mathematics lesson. However, some of them have had an argument about a loan that one had taken out to purchase an item. The interest on the loan is due, but the interest rate is very high and amounts to 100% of the loan itself. This discussion energises the participation of several students as they compare their own families' economic situations. After these discussions, some questions emerge, such as why it is that in some cases people end up paying up to twice the cost of the purchased item.

Such questions were inevitably related to some of the political, social, and economic situations of the students and their families, as is supported by accounts of the interactions that are typical between teachers and members of the community, and the institutional documents (ID)[1] that the school has (Gutiérrez and Rodríguez, 2015, p. 15). We found that the families of the school community belong to the socioeconomic classification of Stratum 1 and 2,[2] wherein the parents have a non-stable job. As a consequence, several parents are in the informal sector and there is a medium level of unemployment.

Regarding family monthly income, this was estimated to be around the equivalent of the current legal minimum wage (616,000 Colombian Pesos for the year 2014)[3] or in some cases even less. Furthermore, in the community, there is evidence of problematic situations related to the sale and consumption of illegal substances, safety issues, domestic violence, forced displacement, poor waste management, gang violence, teen pregnancy, school dropouts, and prostitution (Gutiérrez and Rodríguez, 2015, p. 17). However, at the same time it is also important to highlight that in this community there are young people with a strong commitment to studying, a spirit of solidarity with their families, and hope and aspirations for their future.

The school had established agreements with public and private institutions to promote formative processes with the students in different social spaces such as cultural houses, football schools, health institutions, among others. Regarding the academic field, the mathematics curriculum is divided into four terms for the school year. In this curriculum, we find Ministry of Education (MEN) guidelines about competence with respect to the different components of mathematics. These guidelines are framed around specific topic areas. The common strategy is to evaluate through achievements and indicators, using workshops, guides, tests, and other activities that are

---

1   By Institutional Documents (ID) we mean all the documents that are part of agreements and policies of the school participating in this research. They include the Institutional Educational Project, the mathematics syllabus, the improvement plan, and a document that analyses the context. For ethical reasons we have decided not to reveal the name of the school; for the purpose of this research, we will use ID to refer to these documents (Gutiérrez and Rodríguez, 2015, p. 15).
2   Within this Colombian official stratification, these levels mean working-class people with low income, but not the lowest.
3   For the year 2014, 1 USD = 2,052.52 Colombian pesos. The equivalent of the legal minimum wage was 300,041 USD.

undertaken in the mathematics class (Gutiérrez and Rodríguez, 2015, p. 16).

The above situation, where students expressed their concerns about the credit system, caught our attention as teachers and researchers, as it created the possibility of negotiating exciting and unusual possibilities for subsequent lessons. In addition, one of us was taking a Ministry of Education course about the economic formalisation of education.

Without hesitation, this became the starting point for the construction of our landscape of investigation "Let's Go Shopping", where consumer lending was the aspect to be modelled mathematically.[4] We focussed on the perspective of critical mathematics education by considering some political and social aspects, and we followed the approach of such authors as Skovsmose (1990, 2012), Valero (2006), Blomhøj (2009), among others.

On the research side, our first step was to consider the backgrounds, intentions, and foregrounds of our students, taking into account what happened in the classroom as a result of the discussion of the money loan. Due to this, we moved on to study consumer loans that their families had, especially the one that is linked to the public utility service of electricity supply. We projected a landscape of investigation that allowed us to identify the challenges and possibilities of mathematical modelling in the form of citizenship education through the analysis of student discussions.

Our intention was to establish a learning milieu of mathematical modelling, which creates the possibility of producing discussions about the students' life-worlds, as stated by Barbosa (2006, 2007) and Lerman (2001). A process of negotiation with the students led us to create nine scenes that constitute the landscape of investigation entitled "Let's Go Shopping".

---

4 "Let's Go Shopping" is part of the dissertation *Modelling Your Finances*, a research scenario about education in economy and finances from the socio-critical perspective of mathematical modelling (Gutiérrez and Rodríguez, 2015). It was developed to obtain the degree of Master's in Education with emphasis on mathematics from the Universidad Distrital Francisco José de Caldas (Bogotá, Colombia) and received the guidance of Francisco Javier Camelo Bustos. Reading and comments by Ole Skovsmose.

Scene 1: Tell Me What You Need, and I Will Tell You How to Get It!

The mathematics class started to move towards a different approach from that which was customary. The students were invited to leave their traditional classroom and move to an auditorium to watch a video and to talk about it. The smiles on their faces reflected that they enjoyed what was happening—something that was not common in their previous mathematics lessons. With this in mind, we started the proposal.

The dialogue between the students and us as teachers encouraged the possibility of establishing participation agreements inside and outside of the classroom. We hoped that the blog "Modelling Your Finances"[5] could promote a place for interactions different from that which was common in the regular classroom, and that it would also foster written discussions.

Our students inhabit a world that is highly visual, so for this activity they watched a short video called "Let's Go Shopping" that had been collated by us. In this video, a young 17-year-old girl explores the possibility of opening a savings account. At the end of the video, there is a series of short fragments of advertisements that show people's reactions on acquiring a new product, and several invitations from the financial market relating to consumer lending.

The video caught the attention of all the students, eliciting reactions and comments from them. For example, Jeisson[6] mentioned that "the video invites us to the culture of consumption and the benefits that they offer us for it" while another student, Marlon, expresses that "in December there is joy and in February we are doomed"[7] (Gutiérrez and Rodríguez, 2015, p. 54). The previous statements indicate that some

---

5   The blog "Modelling Your Finances" (http://modelandotusfianzas.blogspot.com) is an online resource that we created to instigate participation and dialogue among the students; through the forums, we collected their perceptions in writing about what is normally done in mathematics classes. It is important to recognise that the students were familiarised with the use of computer skills for interdisciplinary projects.
6   Pseudonyms have been used instead of the real students' names.
7   The original expression said "paila", which is a colloquial expression that denotes a situation that will have a bad result.

students and their families constitute part of a community that actively makes considerable use of consumer lending.

After watching the video, hoping that we could adapt that situation to a classroom activity, the students were organised into groups of four, and they distributed between them the roles of an imagined family from their neighbourhood.

However, during this activity, some students decided to assume the roles of families in socio-economic conditions higher than the ones they came from. As a result, some of the roles selected did not correspond to those in their own neighbourhood. Only three out of the ten groups undertook a reflection based on the economic capacity of their own parents and how they managed to cover their basic needs This information is relevant to what will come in the next scene. Some students who used their own family contexts decided to use the internet on their mobile phones as a way of working with real data. This brought about strong observations to share with the whole classroom, as expressed by Dory, who worked with Jeisson:

> There is not going to be much money left from our parents' salary because they will only be working to pay for our clothes, food, and all that. Our parents should earn a little bit more than the minimum wage, in my opinion (Rodríguez and Moreno, 2015, p. 62, our translation).

Internet access became a tool for the students; it made other sources of information that could help them during the activities accessible. In this case, they discovered the salary that their parents earned, which was previously unknown to them. Thus, some mediation between the contributions of the teachers and the considerations of the students with respect to the knowledge involved in the investigations has taken place. This brings about an insight that goes way beyond mathematics and connects their schooling to their own socio-cultural environment.

The students became involved in an exercise of trust and autonomy by choosing the situations that allowed them to broaden their ideas about their socio-economic contexts. Nonetheless, at this point we realised that neither we nor our students had sufficient experience in creating and executing an educational approach like this one.

Scene 2: Going Shopping Is Amazing!

Once again, the class started with a lot of chatter, which is normal when students are asked to move to another room. This time, the students were asked to go to the computer room because of the opportunities that this space offered for accessing information on the internet.

Our intention for this stage was to provide the students with the opportunity to question themselves about: (a) the items that they wish to buy (branded clothes, the latest model car, a mansion, a private jet, and other luxury things); (b) the reason why all these items are only dreams (high cost of acquisition); and (c) the payment method that they would use to get them (bank loans or credit cards). With the purpose of having a record of their discussions and reflections, we asked them to access the blog and answer some questions in groups. To answer the questions, the students needed to use their own experiences with their families, like in Scene 1. Analysing the situation, they concluded that it was very unlikely that they would ever be able to afford these luxurious items using cash, so they would need to use consumer lending in order to buy them.

When we talked with the students, we recognised the lack of knowledge about consumer lending that they had, and we allowed them to spend some time finding out about the topic. To ensure that all students had a similar experience, we asked them to do a PowerPoint presentation of approximately five minutes. Here, they could present all the relevant information about the selected payment method, and also explain the reason why that method was chosen.

During this time, both students and staff noticed an evident change in the mathematics class. The scene generated an inflection point that challenged the usual routine of consistency on the teacher's explanations and then solving some exercises. This is one consequence not only of the change of spaces, but also of having access to the internet as a pedagogical tool.

Scene 3: Selecting!

For this scene, we were located in the school audio-visual resources room. This is the place where the students were asked to prepare their presentations of the information they researched in Scene 2. Students found four types of commonly used consumer lending. Among them was one linked to the payment of the electricity bill.[8] This is the payment on which the whole landscape of investigation was focussed, and we sensed an uncomfortable atmosphere in the classroom that day. The positive and joyful environment that we had seen in the first two scenes seemed to be fading away. The students seemed distracted and not focussed on the objective that we were proposing.

Time kept running out, and the class quickly ended, but it left us researchers thinking about what we were doing. We imagined that we had presented a routine-like exercise of gathering information from the internet. If we look closely at what did happen, the silence that was experienced when the students were asked to come up with questions could have been because each student was only interested in what they were doing in their own group. We can see some evidence of that in the following discussion:

> *Julian*: Well, the documents required for this loan can be identified here [pointing to the image projected in the video beam]. If the person receives a salary, a pension, or is an independent worker, they need to have: an enlarged copy of the ID, a certificate of incomes and deductions, a work certificate, a copy of the last salary payment slip, a copy of the income tax certificate from the previous year, a certificate from the chamber of commerce dated not less than 30 days before, a copy of bank statements for their savings or checking account over the last few months, financial statements—those are some of the requirements to get a loan.
> *Teacher:* OK, do you have any questions about any of the terminology that is being used there, any of the conditions they have talked about… [Silence]. Nothing? Is everything clear? Is everything understood? [Silence]
> *Steven:* Questions, whoever…. [Students laugh]
> (Gutiérrez and Rodríguez, 2015, p. 67, our translation)

---

8   This type of consumer lending has been implemented as an efficient mechanism for the low-income population in Bogotá so they can have access to the banking system. The requirements for this are easy to meet and people whose income is as low as half of the current legal minimum salary (around 150 USD for 2014) can also acquire a credit card. These people could be independent workers, employees, or retired people, without the need for a co-signer. The credit study only requires 48 hours for approval, and the person does not need to have a credit score or be a homeowner. The payments are charged to the monthly electric bill (Crédito fácil Codensa, 2020).

Based on what was said before, we needed to reflect on what was taking place. At this point, we felt the pressure of not knowing whether we were actually achieving the aim of creating a different type of mathematics education. Some questions emerged, such as: what was the purpose of the presentations done by the students? Why did we ask the majority of the students to develop a similar activity? Maybe we did in order to ease the monitoring of the activity? Or maybe our experience as teachers called us to go back to our comfort zones? At this moment, the tension rose because we did not know if the students would be able to display their real interests in front of the class. After all, from the outset the targeted objectives had been proposed by us instead of them.

Then we remembered that idiomatic expression, "the darkest hour is just before dawn". Our doubts can also be sources of clarity in the opportunity to reflect on our decisions and actions; they show us a way to advance in our work towards a learning milieu of type 6, as is proposed by Skovsmose (2012) (bear in mind that learning milieu of type 6 is characterised by reference to real-life situations). This caused us to face a great challenge: to allow the students to move beyond explicit instructions and beyond their constant need for the approval of the teachers. On the contrary, the objective of the landscape was to guide the students to work in a cooperative process of constant negotiation, and in that way, to construct an ideal landscape.

By rethinking our purposes, we recognised, as Skovsmose (2012) proposes, that during the exploration of a landscape of investigation it is possible to move between different learning milieus. The idea is that it is not advisable to remain in one particular milieu, or to continue in one that does not contribute to the intentions of the researchers or the students.

Scene 4: Choosing!

A new lesson started, and with it the opportunity to start again with the objective of changing the style of lesson. This new space gave us an opportunity to build, along with our students, a tool that was meaningful and allowed them to interpret and transform their lives and their social contexts.

Following that plan, students were instructed to enter the blog and the forum entitled "Finally, What Payment Method Have You Decided to Use for Your Shopping?" In this forum, the students participated by expressing their decisions based on the presentations and their research. They also needed to clarify whether they had changed their mind about the payment method since the beginning of the activity. Here is an extract of what one of the students wrote: "we decided to use the Codensa card,[9] because it gives us many benefits, and now we can do cash loans. This is easier to obtain and manage" (Gutiérrez and Rodríguez, 2015, p. 71).

Thus, each action and decision developed by the students was recorded both in the blog and in the work folder, allowing them to identify the importance of knowing the process of applying for a loan, including the way in which the amount granted by a financial institution is defined. This last aspect became the generator of the guidelines that shape Scene 5.

It is important to recognise that even the most difficult moments can provide opportunities for growth. Our experience made us realise that the presentations did not emerge from a personal need felt by our students. Nevertheless, the presentations gave them access to much information, and this information changed some of the decisions that the groups had originally made.

Scene 5: What Do I Need?

In the continuous search to know and understand the process of applying for a consumer loan, the students filled in the forms required for this application. Consequently, they started to recognise the relationship between the income of the person requesting the consumer loan and the amount of money that this person might be given. We can see evidence of that in the following remark by Julian:

> Here! This is when we did it wrong because we put the salary of Stiven and his income that is 100,000 COP (Colombian Pesos) per week. If we are going to request the loan in Codensa, they said that they can give you

---

9   The card or loan Codensa, mentioned by the students, refers to consumer lending that is linked to the electric service company explained on the previous page.

up to four times your income. The top amount that they will loan him is 2,000,000 COP. That means that he wouldn't be able to afford anything[10] (Gutiérrez and Rodríguez, 2015, p. 77, our translation).

The students' way of working started to please us as, without needing too much guidance, they managed to establish for themselves a cause-effect relationship between the variables of the mathematical model applied for the approval of consumer lending (Barbosa, 2006).

At this point, the tensions that we faced in the construction of the landscape of investigation eased slightly, and we were able to view the negotiation with the students with optimism. Following this thought, we could create the possibility of fostering a learning milieu in which mathematics became a tool that allowed them to understand the decisions that are taken in society and in certain situations.

As the students continued exploring the topic, they started to reflect on what happened in their own families, and related it to the work they had been developing in class.

Days like this allowed us as researchers to open up a space for reflection, a space in which we found evidence of the students' progress towards autonomy. They oriented their interests by rooting them within their own concerns, and not within what the teachers gave them. This progress increased our confidence, and made us believe that it would be easier to advance.

Scene 6: And My Family? Are They Going to Be Indebted?

When students embarked on an examination of how their families did their shopping, the dull walls of the classroom—that seem to be designed to exclude or protect the students from the dynamics of the social world of adults (though eventually, they are going to face that world)—opened up for them. They came to recognise what happens outside the walls of the classroom and inside the walls of their homes, with respect to financial difficulties.

---

10   The shopping ambitions of the students included the acquisition of items and services with high cost such as branded clothes, last model cars, mansions, and private jets (among other things).

This recognition becomes very important because it brings back into the classroom information that comes from real-life contexts. It is modelled through mathematics, just as we saw in the forum blog in the intervention that Allison did: which are the payment means related to consumer lending that your family is currently using?

> The head of the home has requested a loan from the bank 'Finamerica'. At the beginning they gave her some advice; she needed the loan for a family meeting that could not be postponed. The request was for 2,000,000 COP and it was approved. In order to get it, she went to the bank, and they only approved 1,150,000 COP because it was her first time and at the end, they only gave her 1,007,550 COP, saying that the rest of the money was a fee for the documentation (Gutiérrez and Rodríguez, 2015, p. 78, our translation).

At the same time, we saw that, despite the uncertainty which emerged through the development of our landscape of investigation, the work should not be forced or simulated. It should be inspired by the students' own interest in the situation and the living experience that it offers them.

The previous situation allowed the students to establish a relationship between the economic hardships that exist in their homes, caused by the use of consumer lending, and the fact that the money that should cover basic needs is used for paying debts. A student named Angie mentioned that "the situation in some homes is sometimes a bit complicated because there are many debts from loans and credit cards" (Gutiérrez and Rodríguez, 2015, p. 81, our translation).

The conversations among some of our students left us with a lot of questions about how to move forward with the activity. This situation drove the students to seek the approval of the teachers who were traditionally seen as the authority in terms of knowledge. However, it takes time for students to become truly autonomous, as Steven commented: "it takes her forever, and at the end, she provides no explanation" (Gutiérrez and Rodríguez, 2015, p. 79, our translation).

Based on what they said, we recognised almost immediately that our students still looked to us for *our* approval about what *they* thought and did. This experience of security helped them to continue with their work. The discomfort that Steven expressed is common among students, and Julian mentioned: "I get desperate because I start to look for help using other means. I can get support from one of my classmates, but in the end it is the teacher who knows. That will give me the security that it is solid" (Gutiérrez and Rodríguez, 2015, p. 80, our translation).

At that moment, we faced a challenge in the learning milieu that we wanted—and still want—to foster in the mathematics classroom. Our interest is that, in this landscape, students work autonomously, and the teacher is no longer the person that possesses absolute knowledge. Consequently, we shifted our attention towards the possibility of opening up spaces of democratic participation where the knowledge of the students is also recognised.

Scene 7: A New Look to Your Shopping

The time came when students started questioning the reality of the credit system and how much people end up paying when they use consumer lending. This concern drove students to research more about how companies determined the number of payments and the value of them.

At this stage, the internet became a very important part of the landscape, because students used it to interact with a virtual simulator. The simulator allowed them to recognise that the value to be paid depends directly on the number of instalments of the loan, something they had previously been unaware of. The students were asked to register this information in their folders (as seen in Fig. 1 and Fig. 2). In this activity, mathematics continued to be the vehicle in the process of investigation.

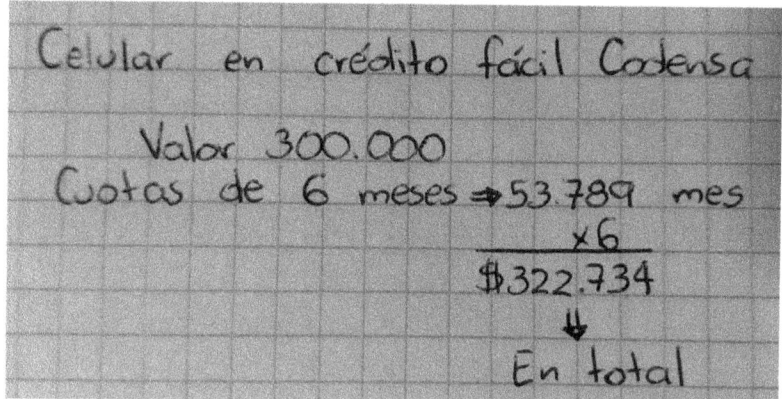

Fig. 1. Registration in the work folders of the students (Gutiérrez and Rodríguez, 2015).

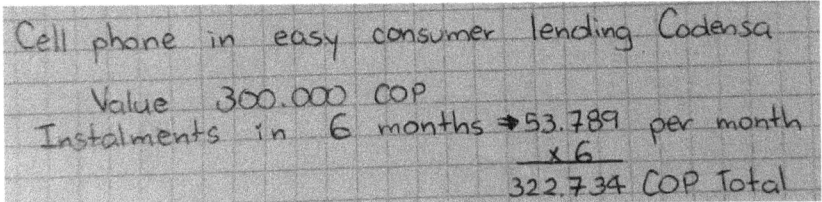

Fig. 2. Translation of the work folder.

The exploration of the landscape then moved forward, not only at the level of our pedagogical practices, but also at the level of the objectives that we had set initially. We found that the students had developed a better comprehension of some elements of the mathematical model and the way it functions. Nevertheless, the students were still not able to fully explain the process of calculating the value of each instalment, as we can see in the following observation made by Julian:

> The company Codensa must earn some money for each payment. For the customer it is more expensive because they have to pay more than what they asked. This process generally happens without the customer being fully aware of the extra money that they are paying for the item they purchased (Gutiérrez and Rodríguez, 2015, p. 86, our translation).

Scene 8: Now What Do We Do?

On this day, two of the groups that had worked on the same credit system interacted with each other to answer questions. Their concerns were about the model employed to obtain the interest for the instalments, and their relationship with the existence of exceeding value. This value is charged to the customer without them really knowing where it comes from. This can be seen in the following conversation:

> *Karen*: Julian, look here! What else can I do? What do you think is missing?... 32,104, we are missing those 32 that they are charging us extra, but we do not know why. Look here! There is a 28%. I do not know where that came from.
> *Julian*: We also found the 28 point something, and we calculated the formula of percentage and that could not be the price that they charged you. I would say that is 2.1276%.
> *Karen*: And where does it come from?

Julian: From the rate... the interest rate, but I do not know, I do not know if what I am doing is right.
Karen: No, and I do not know what to do...
(Gutiérrez and Rodríguez, 2015, p. 87, our translation)

Taking into account Julian's confusion about whether or not his procedure was correct, we had the opportunity to reflect on how his insecurity did not necessarily mean a defeat for us teachers. Actually, it gave us evidence about the process of questioning a specific model of the credit system. Unlike the previous scenes in which we were uncertain about the upcoming events, this experience allowed us to direct the work of our students more effectively. In this case, it showed that, in this space, students can reflect about the functioning of consumer lending.

At the same time, the previous situations let us understand that, through reflection about the mathematical modelling, reflexive and technical discussions tend to emerge according to Barbosa (2006). These discussions established a relationship between the interest rate expenses and the number of payment instalments that the total debt is divided into. In this landscape of investigation with all its ups and downs, one student—Lorena, who had been working with Karen—had a critical position towards the functioning of the mathematical model, and she expressed her opinion as we can see in the following extract from her work folder:

> Despite many attempts and mathematical operations, there is no way to get a specific result. We did the calculations over and over again without knowing where all that amount of money they were charging came from. This means that the companies are stealing from the people who have to request consumer lending. The worst thing is that the people do not realise it. They just accept the terms and pay for the money, which they should not do in the first place. This exercise helped us a lot, because it made us realise what is in store for us (Gutiérrez and Rodríguez, 2015, p. 90, our translation).

At this point, we can see that the objective of the research scenario with respect to the process of modelling is properly oriented. We achieved this despite the fears, uncertainties, and tensions that were experienced during the whole process, not only by the students but also by us, the teachers and researchers.

Other situations that are worth highlighting here are the interactions among students. Karen, for example, mentions that she approached Julian's group because he normally "did well in mathematics"

(Gutiérrez and Rodríguez, 2015, p. 89, our translation). This statement shows us the empowerment that a person can experience when he or she has knowledge in this area. It establishes a power relationship that we perceive in this case as a challenge. Considering that, with respect to this landscape of investigation, it is fundamental to recognise how dialogue and teamwork contribute to citizenship education, as formulated by Silva and Kato (2012).

Based on what happened, we felt that it had been a good day for all of us. We believed that, one way or another, each participant had achieved, at their own pace, a socio-critical perspective on the class and their own reality.

Scene 9: The History of Things

The critical reflections that started to take place in the classroom made it possible for personal opinions about the credit system to emerge. This was a good opportunity to create dialogue. In this way, our students started sharing their points of view that were supported by their own explorations of the landscape of investigation. They also recognised the possibility of relating these ideas to the situations of their own families.

The satisfaction that we experienced when we saw the path that we had taken was comforting. This satisfaction allowed us to consider a family simulation, like the one that happened in Scene 1, as being a functioning educational approach. It showed that students can empathise with their families—in more colloquial terms, they can put themselves in their families' shoes. They came to consider something that they had never thought of, neither in the classroom nor in their homes (Gutiérrez and Rodríguez, 2015, p. 94).

Additionally, the milieu that was created allowed us to retry, at some point, the processes of negotiation with the students. Through the creation of a forum, the work students developed in the classroom could be shared with families and other members of the community. This means that the landscape of investigation could open the doors of the classroom towards problematic socio-economic situations.

Nevertheless, today we have to recognise that the institutional pressure within the school has started to intensify. Due to decisions taken at the administrative level of the school with respect to the school calendar, and specific instructions for the academic end of the year, we were unable to do what we had wanted to do at the beginning of the project with respect to the creation of the forum. In the following fragment of an InterView,[11] we can see some indications of this:

> *Teacher*: We have talked about a forum that was going to be established on a Saturday because there were some time issues and other institutional activities, but we couldn't make it happen. Would it have been good to do it?
> *Miguel*: Yes! I think that we could have done that reflection with the parents, the one that we did ourselves.
> *Lorena*: I think it would have been useful because some adults do not know where all of this comes from. They just pay. But all this information would have been very useful.
> *Julian:* Well, it was very interesting and very nice to make that forum. We could have brought our relatives—parents, cousins, uncles—so they could get more information about this important topic.
> *Jeisson*: I would have taken my whole family, because many of them are not aware of this.
> (Gutiérrez and Rodríguez, 2015, p. 97, our translation)

The InterView allows us to recognise the intention of moving beyond the walls of the classroom. We tried to get the families and other people from the community involved in the process of mathematical modelling from a socio-critical perspective, but that turned out to be too challenging. In this way, the reflexive discussion among the students would not be able to widen the impact of this landscape of investigation. Nevertheless, it could create greater engagement of society through the students. This observation is supported by Silva and Kato (2012).

The time to end the class had arrived, and also the time to conclude the landscape of investigation. There was only time left to thank our students for allowing us to learn with them, to recognise their interests, and to see how they gave new meaning to their realities through mathematics. Consequently, we can acknowledge that the tensions, uncertainties, and fears were worth it, just like Julian mentions in one of the InterViews:

> It is important to change the space you're in; we were locked down in a classroom for about eight hours and we continued with the same monotony... we moved to the library, the audio-visual room, the computer room—when we did that we could clear our minds for a moment and we were more relaxed; we knew that we had a problem to solve and we

---

11 An InterView is where knowledge is built through the interaction between the interviewer and the interviewee. See Kvale (1996).

went to solve it. We had the opportunity of using the internet and we could find out more about what we had to do. It was important to change spaces because if we did not have all the information to support what we thought, we would not be able to change our minds (Gutiérrez and Rodríguez, 2015, p. 101, our translation).

10. Reflections

The set-up of the landscape "Let's Go Shopping" gave us numerous possibilities and some challenges due to our aim to create a learning milieu of type 6 (Skovsmose, 2012). We considered that the interactions among the students should justify each action and each step forward in the investigative process.

However, our lack of experience and training constantly pushed us to plan and over-specify the instructions to guide the students' work. As a consequence, to a stronger extent than we had hoped we were leaving aside the intentions, backgrounds, and foregrounds during the negotiation process, all of which might give meaning to working with the landscape of investigation.

Bringing together all the classroom participants made it possible to establish reflections about the social aspects embedded in the project, with respect to the use of consumer lending and its impact on the families' finances.

Researching the landscape of investigation made it possible to take the mathematics lesson to other locations than the classroom. In this way, we reinforced the use of technological tools that helped students undertake research beyond the use of texts and information provided by the teacher.

As part of the exploration of the landscape of investigation, we observed a change in the students' views. This seemed to be due to moments of reflection about social aspects that were not considered before in the mathematics class. Thus, the students' discussions allowed us to identify understandings and insights about their role in society and the reality in which their families are immersed.

At the beginning of this project, many expectations, fears, and uncertainties emerged because it involved a negotiation process of the landscape of investigation with the students. Some of these expectations gave the project a new meaning, both for us and for the students themselves. This provided us with the starting point of a path that leads us to further explorations.

References

Barbosa, J. (2006). Mathematical modelling in classroom: A socio-critical and discursive perspective. In C. Haines, P. Galbraith & B. Blum, *Mathematical modelling: Education and economics* (pp. 293–301). Horwood Publishing.

Barbosa, J. (2007). Mathematical modelling and parallel discussions. [Paper presentation]. In D. Pitta-Pantazi & G. Philippou, *Working Group 13: Modelling and applications*, Larnaca, Cyprus (pp. 2101–2109). http://www.mathematik.tu-dortmund.de/~erme/CERME5b/WG13.pdf

Blomhøj, M. (2009). Different perspectives in research on the teaching of learning mathematical modelling. In M. Blomhøj & S. Carreira (Eds), *Mathematical application and modelling in the teaching and learning of mathematics. Proceedings from Topic Study Group 21* (pp. 1–18). https://rucforsk.ruc.dk/ws/portalfiles/portal/3820977/IMFUFA_461.pdf#page=6

Codensa. (2020). ¿Qué es Crédito Fácil Codensa? Crédito Fácil Codensa. https://www.creditofacilcodensa.com/ciclo-de-vida/acerca-de-cfc

Gutiérrez, F. A. G. & Rodríguez, Y. C. R. (2015). *Modelando tus finanzas. Un escenario de aprendizaje sobre educación económica y financiera desde la perspectiva socio-crítica de la modelación matemática*. Master's thesis. Universidad Distrital Francisco José de Caldas.

Kvale, S. (1996). *InterViews: An introduction to qualitative research interviewing*. Sage Publications.

Lerman, S. (2001). Cultural, discursive psychology: A sociocultural approach to studying the teaching and learning of mathematics. *Educational Studies in Mathematics, 46*(1–3), 87–113. https://doi.org/10.1023/A:1014031004832

Mejía, D. (2010). *Estrategia nacional de educación económica y financiera en Colombia*. Docplayer. http://docplayer.es/21797244-Estrategia-nacional-de-educacion-economica-y-financiera-en-colombia.html

Silva, C. & Kato, L. (2012). Quais elementos caracterizam uma atividade de modelagem matemática na perspectiva sociocrítica. *Bolema, 26*(43), 817–838.

Skovsmose, O. (1994). *Towards a philosophy of critical mathematics education*. Kluwer Academic Publishers.

Skovsmose, O. (2012). Escenarios de investigación. In P. Valero & O. Skovsmose (Eds), *Educación matemática crítica: Una visión sociopolítica del aprendizaje y la enseñanza de las matemáticas* (pp. 109–130). Centro de Investigación y Formación en Educación, Universidad de los Andes.

Valero, P. (2006). ¿De carne y hueso? La vida social y política de la competencia matemática. *Memorias de Foro Educativo Nacional*. Ministerio de Educación Nacional.

3. Media and Racism

Reginaldo Ramos de Britto

> In this paper, I address the theme of racism in the mathematics classroom in order to investigate the invisibility of black people in printed media. To do this, I describe a pedagogical strategy called the Social Research Group (SRG), which largely originates from critical mathematics education. SRGs are research groups formed by students in basic education who develop thematic investigations. In this text, we describe how the theme of racism was problematised through one of the landscapes of investigation, built on the theme of the visibility of black characters in national magazines. This scenario, in addition to enabling reflection on an important topic for Brazilian society—racial democracy—served to promote the idea that mathematics not only colonises various social practices, but can also be an instrument that helps us to reveal social asymmetries.

In this chapter, I discuss racism in printed media in national circulation magazines in Brazil, through a teaching practice which I term Social Research Group (SRG). It is about a landscape of investigation built in the mathematics classrooms where I work.

The SRGs are groups formed by public school students from elementary education and the practice aims to develop thematic investigations. The pedagogical proposal of the SRG is the result of a prior proposal entitled Mathematics Education and Democracy, which since the beginning of the 2000s has had the goal of taking relevant social themes into the mathematics classroom.

The SRG has its own organised structure that revolves around the proposition of a landscape of investigation, and aims primarily for students and researchers to acquire and develop the competence of mathemacy.[1] Although there are multiple meanings given to the expression "pedagogical projects", we have to consider that the majority of proposals that use this name have in common the intention of characterising and describing an action that tries to untangle itself from the usual and traditional way of developing the teaching-learning process.

The SRG has the intention of humanising the mathematics classroom, in the sense of transforming it into an environment able to contribute to the discussion of themes such as the history of the neighbourhood; teenage pregnancy; inclusion of students with disabilities; the students' musical tastes; religious positionings; the electoral process and knowledge of the school community; social inequalities, and media and racism. These themes were discussed in the landscapes of investigations throughout the years in which the project has been developed.

At the schools where I am based, the strategy of work through projects has not yet become a reality at the curriculum level, and the SRG, generally, is confined to the eighth- and ninth-grade mathematics classes of a municipal school, and the second and third grades of a state high school. When the nature of the research does not require us to step outside the school premises, the research is carried out in the SRG rooms (Fig. 1).

Fig. 1. SRG research classroom in one of the schools.

---

1   Landscapes of investigations and mathemacy are concepts described by Skovsmose (2008).

The idea of putting together a research group consisting of basic education students was inspired by the social fragilities in the communities where the schools are located.

This social context favoured themes such as racism and social inequalities to be approached in our mathematics classes. I was concerned by the fact that important social matters, relevant to democracy, were not being addressed in my classroom. My reading of *Pedagogy of the Oppressed* (1987) and especially *Educação e Mudança* (1983), both by Paulo Freire, certainly contributed and helped to promote a change in my professional trajectory, awakening me to the importance of the "political commitment" of the education professional. Racism, and social and racial inequalities, were certainly the most discussed topics in this trajectory. This is largely because the schools in which I work, in Juiz de Fora, Brazil, are located in neighbourhoods with a significantly large black population. Table 1, formulated using information from the Brazilian Institute of Geography and Statistics (IBGE), with contributions from Vanísio Luiz da Silva, indicates the respective presence of the black population in the neighbourhoods of the schools in which the projects are located.

Table 1 presents statistics for some of the neighbourhoods within the school district. The first line shows the population data from the city of Juiz de Fora, regarding its ethnic composition. In the city of Juiz de Fora, 43% of the population are black people.[2] The ethnic profile of the students from these schools reflects that the majority are black students.

---

2   Officially, there are two categories (black and brown) used by the Brazilian government, and those will appear in official government data. As a political statement, I adopt the term "black" regardless of race mixture, reducing the government categories to only one: black people.

Table 1. Population according to colour/race in the neighbourhoods in which the Social Research Group (SRG) schools are located. Data from IBGE Demographic Census 2010. Copyright 2010 by IBGE.

REGION (1)	TOTAL	WHITE	BLACK	ASIAN	BROWN	NATIVE	NOT/DECL	BLACK
JUIZ DE FORA	728602	409355	105425	4929	208098	792	3	43%
REGION (2)	TOTAL	WHITE	BLACK	ASIAN	BROWN	IND	ND	BLACK
IPIRANGA	16045	7248	3798	166	4819	14	0	8617 (54%)
SAGRADO COR.	2716	801	546	76	1292	1	0	68%
SANTA EFIGÊNIA	7669	2722	1753	55	3135	4	0	64%
SÃO GERALDO	4227	1784	991	29	1419	4	0	57%
REGION (3)	TOTAL	WHITE	BLACK	ASIAN	BROWN	NATIVE	NOT/DECL	BLACK
SÃO BENEDITO	14693	6041	3375	214	5012	51	0	8387 (57%)
SÃO BERNARDO	3649	2566	340	7	734	2	0	29%
VITORINO BRAGA	4205	2853	399	23	922	8	0	31%

1. Racism as an Erosion of Democracy

If we think in terms of geology, erosion is a corrosive process of degradation. In this scenario, there always exists a cause, which could be as natural as the rain and the wind; it could also be a result of human action. Despite the climatic changes provoked by human action, bigger and more harmful each time, it is possible to comprehend erosion as a process of nature and, therefore, as unintended. When provoked by humans, however, erosion is always conditioned to processes of production, which promote the degradation and exhaustion of the planet's natural resources.

However, erosion can also refer to degenerative processes in the social relations between a group of people, such as the relationships between teachers and students, or between students and their peers.

In my point of view, erosion represents, primarily, fissures in democratic relations, resulting in social impacts that produce power asymmetries. Considering the asymmetry of racial relations, racism is an erosion that, when studied and discussed, allows us to unveil the myth of racial democracy in Brazil.

There is much statistical data and research that reveals disparities between white and black people that cannot only be assigned to socioeconomic gaps. In particular, the data that has attracted my attention concerns the average years of study completed by white and black people:

> Between 1995 and 2015, the adult white population with 12 or more years of study came to double, from 12.5% to 25.9%. In the same time frame, the black population with 12 or more years of study goes from unbelievable 3.3% to 12%, a raise of almost 4 times more, but that does not hide the fact that the black population only now reaches the degree reached by the white population twenty years ago (IPEA, 2017, p. 2)[3].

In 2016, the illiteracy rate was 9.9% amongst black people and 4.2% amongst white people. Out of the total of 1835 children between 5 and 7 years old that worked in 2016—despite child labour being

---

3 Available at http://www.ipea.gov.br/portal/images/stories/PDFs/170306_retrato_das_desigualdades_de_genero_raca.pdf

prohibited—around 64% were black or brown.[4] The labour market offers different levels of salary to white, black and brown people: R$ 2814.00 to white people, according to the 2017 income average; R$ 1606.00 to brown people and R$ 1570.00 to black people.[5] Although the disparities and racial differences have an economic component, they are structural and represent the erosion of the racial democracy. Critical mathematics education, with its concerns over democracy, is a favourable environment within mathematics education in which to reflect on these issues.

Besides this, we inherited a school built upon a euro-centric epistemological basis that "persists and influences the school curriculum, even in a subject supposedly neutral as mathematics" (Frankenstein and Powell, 1997, p. 2), which reinforces the need for researchers, as well as the curriculum in mathematics education, to address the theme of racism.

2. The Whitening Ideology in Brazil

The *whitening ideology* is the name given to the historical process that Brazil used to end the so-called "negro problem" after slavery ceased. It is about the theories "accepted by the majority of the national elite in the period between 1889 and 1914 (...) based on the assumption of white supremacy" (Skidmore, 2012, pp. 110–111, my translation).

The whitening speeches helped to build an *"imaginary"*[6] which constructs our understanding of racial relationships in Brazil, granting a social position of inferiority to the black people.

Those speeches were influenced by racist theories produced in the 19th century by such characters as Arthur Gobineau (1816–1882), an aristocrat and scientist who took a pessimistic stance on the mixing of races in Brazil. He was the "most important inventor of the Aryan myth of racial content [...] [who] claimed that a certain mix of Aryans with

---

4   Available at https://agenciadenoticias.ibge.gov.br/agencia-noticias/2012-agencia-de-noticias/noticias/21206-ibge-mostra-as-cores-da-desigualdade
5   Idem.
6   An "imaginary" consists of a set of values and symbols through which people imagine their social world. Derived from Sartre in 1940, and used by Durant and others, an "imaginary" is a construct parallel to our experience of reality, through which all of us look for meaning in our everyday lives. This is fed by many of our beliefs, ideas, traditions and preconceptions, as well as our scientific knowledge.

inferior peoples would have been fundamental to unfolding the civilising processes, although uncontrolled crossbreeding was the main cause of civilisation's degeneracy" (Seyferth, 1995, p. 180, my translation).

In Brazil, several intellectuals have reproduced these ideas during the Republic period. They were sociologists, jurists, doctors, engineers, all advocating for the same cause: get rid of the "negro problem" in Brazil.

According to Nina Rodrigues (1862–1906), a doctor from the state of Bahia, "the black race in Brazil, no matter how big were their undeniable services to our civilization [...] must always constitute one of the factors of our inferiority as people".7 To Pierre Denis (apud Domingues, 2004, pp. 51–52), a brazilianist,[8] the black people were "indolent; the work inspires in him a profound horror; and he will only surrender to it forced by hunger or thirst". On the other hand, Oliveira Viana (1883–1951), a jurist, considered that: "Pure blacks can never fully assimilate to Aryan culture, even the highest specimens: their ability to become civilised, their civilizability, does not go beyond imitation, more or less perfect, of the habits and customs of the white man" (apud Domingues, 2004, p. 260).

This historical phenomenon helped to create, here in Brazil, discrimination against black people. The printed media, by preferring to portray white characters and giving black people little attention, have reproduced this ideology.

3. Social Research Group: Media and Racism

In mathematics classrooms, one of our research group tasks is to analyse the visibility of black people in national circulation magazines, an activity we have performed since 2006.[9] The task is entitled "Media and Racism" and the goal is to demonstrate how mathematics can help to unveil the level of participation of black characters in advertising articles from magazines.

---

7   From the book The Africans in Brazil. Available at: http://www.brasiliana.com.br/obras/os-africanos-no-brasil
8   Foreign people that study Brazil.
9   This landscape of investigation had been discussed before this date, but this milestone refers to the oldest research that we have some kind of record of.

This landscape of investigation was developed further in the years of 2009, 2010, 2012, 2013, 2015, 2016, 2017 and 2018. Not all years have records available and the following discussions refer to those landscapes with reliable records.

Furthermore, this investigation represents an intent to demonstrate that mathematics is an important tool for unveiling situations and daily social practices that present erosions of democracy, but remain unnoticed most of the time, present themselves as natural and do not allow us to question them.

One of the main goals of the actions by the Social Research Group is to transform the mathematics classroom into a privileged space to develop "investigations and critical reflections about social themes that are relevant to citizenship" (Britto et al., 2017, p. 383).

Landscapes of investigations like these allow us to understand that mathematics—at the same time that it colonises and structures social practices—and serve to unveil asymmetries of power and fight erosions of democracy, such as racism.

A landscape of investigation "is the one that invites students to formulate questions and seek explanations" (Skovsmose, 2000). In the case analysed in this chapter, the generating question for the investigation formulated by the students was: does the print media offer the same space in advertisements to black and white people?

The following narrative describes what, altogether, occurred during the years that this landscape was being developed.

The students were divided into small groups. Each group received the task of collecting pictures of ordinary characters in national circulation magazines within the period of one week.[10] Each group had to select a specific number of pictures that varied throughout the years. Each picture could only portray one character and the students had to understand the context in which the character was presented. Most of the pictures were collected out of school time and, in some cases, in the research group classroom, where the students looked for the pictures in a pile of magazines that were put together for this purpose.

---

10 In the initial versions of this landscape of investigation there were only pictures of children. Later on, we expanded the reference frame using the term "character" to refer to the representations (or not) of ordinary people, found in diverse activities and situations in the magazines that were searched.

At this point, we suggested to the group that students should not make character choices but only collect photographs. People, in general, mobilise their personal preferences when making choices that, for this reason, are influenced by social, ideological and cultural determinants. The idea was to allow a certain amount of randomness to guide their choices.

Until this phase of the activities, in some of the versions of the landscape, the students received no information about the objectives of the landscape under analysis. As mentioned, not all editions of the project explicitly presented students with an investigative question. We believed that if they knew that the activity was related to racism, it could influence the choice of characters in the pictures.

After the photo collection, the students performed quantitative work, adding the number of pictures collected by the small groups. They determined which, and how many, characters were white as opposed to black. Then they had to describe what we call "the surroundings of the pictures". Just as we thought, the environment plays the same role as the context in linguistics. The environment is what defines "how we see the present situation and how we act in it" and represents "a fundamental component of our understanding of human conduct" (Van Dijk, 2012, p. 21).

In sequence, we asked each group to analyse the situation surrounding each character in the photo, the context, and estimate whether it was "positive" or "negative". This decision was made by "reading" the circumstances surrounding the photo.[11]

For the 2006 edition of the project, there was no general compilation of data from all small groups, so we reproduced the information extracted from the work of one of these groups (Fig. 2 and Fig. 3).[12] Of the total of 41 pictures collected, 36 represented white children and 5 portrayed black children. In addition, of the 36 considered white, 35 (97.2%) were in positive situations and only 1 (2.8%) in a negative situation. Of the

---

11   As the environment is either related to overflow or to what is shown, we considered it appropriate to describe the social context that involved the character portrayed in each photo, as a strategy to understand the outspread of discrimination on the racial imaginary.
12   There is a small adaptation in our text because in 2006 the task was to collect pictures in which the characters were children. At that time, we were interested in observing the visibility of black children in printed magazines.

5 pictures with black characters, 3 (60%) were in situations considered positive and 2 (40%) in situations considered negative.

Número de fotos de crianças Brancas	Positiva (P)	% Positiva (P)	% Negativa (N)
36	Negativa (N) 35 1	97,2%	2,8%

Fig. 2. Data produced by one of the subgroups in the 2006 edition of the project about the degree of visibility of white and black characters.

Número de fotos de crianças Negras	Positiva (P)	% Positiva (P)	% Negativa (N)
5	Negativa (N) 3 2	60%	40%

Fig. 3. Data produced by one of the subgroups in the 2006 edition of the project, about the quality of the participation of the white and black ethnic groups.

In addition to the quantitative work that was performed, this display is really indicative of the inferiority imagery to which we referred. The total number of pictures that depict children identified as white (36) is strongly superior to the number of pictures that depict children identified as black (5). The descriptions of the pictures' surroundings also reveal that the former ethno-racial segment is, proportionally, more often thought to be in positive situations. These asymmetries constitute erosion of racial democracy; publicity articles should be franchised in a way that is ethnically and socially egalitarian to represent that the population of Brazil is made up of 44.2% white people, 46.7% brown people and 8.2% black people.[13]

Some of the subgroups from the project's 2010 version moved on in their investigations. In their reports, one of them identified prejudice and discrimination "not only in the media", claiming that "if a black man walks in a supermarket, a security guard quickly follows him" (sic.) (Fig. 4). This group did not restrict its search to the analysis of racism in the press media, as suggested in the landscape of investigation. The group contributed further by sharing really meaningful personal experiences, as can be observed from the extract shown in Figure 4.

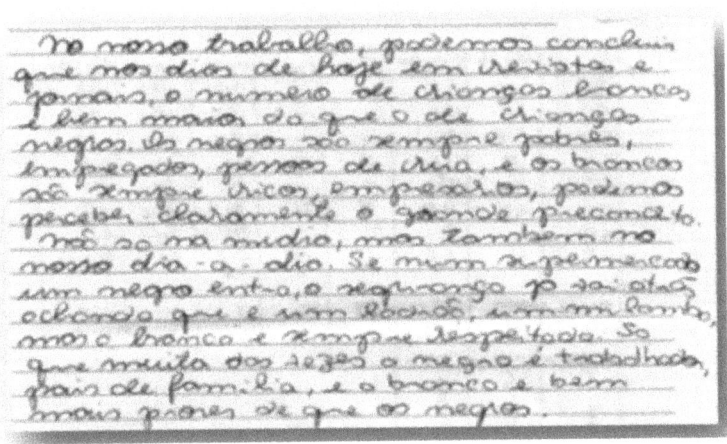

Fig. 4. Subgroup report on the 2010 version of this project.

The text in Figure 4 runs as follows: in this work of ours we can conclude that nowadays, in magazines and journals, the number of white children is much higher than black children. The black people are always poor, housekeepers, homeless and white people are always rich and entrepreneurs. We can clearly perceive the huge prejudice, not only in the media, but also in our day-to-day lives. If a black person walks in a supermarket, the security guard quickly follows him assuming he is

---

13 Available at https://agenciadenoticias.ibge.gov.br/agencia-noticias/2012-agencia-de-noticias/noticias/18282-populacao-chega-a-205-5-milhoes-com-menos-brancos-e-mais-pardos-e-pretos.

a thief, but white people are always respected. But most of the time the black people are hard workers, family-oriented parents and the whites are much worse than the blacks.

The identification of the ethno-racial "type" was also an activity that caused interesting discussions. These can be observed in the extract in Figure 5.

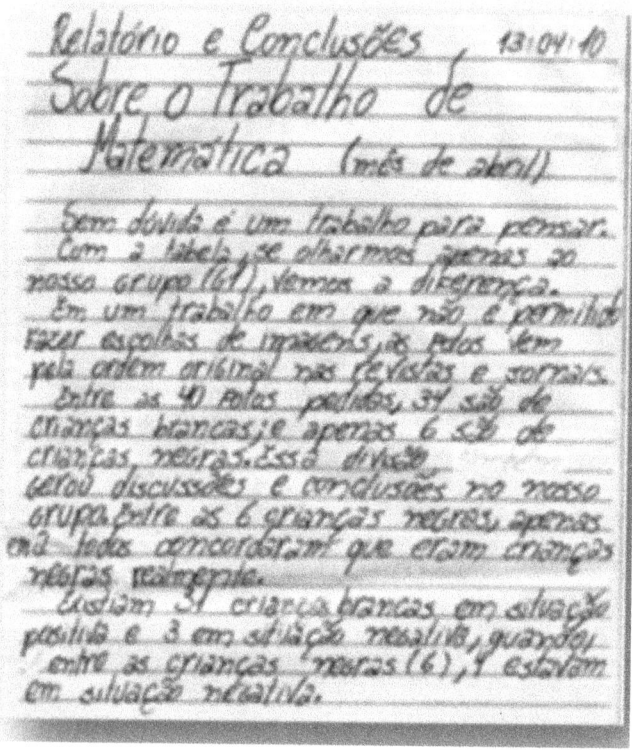

Fig. 5. Report of the subgroup on the 2010 version of this project.[14]

---

14 Report and conclusions: about the Mathematics Assignment (month of April). No doubt this is an assignment created to generate reflections. With the table, if we only look at our group (G1), we can see the difference. For a task where we are not allowed to choose pictures, the pictures are given to us in their original order in the magazines and newspapers. Among the forty (40) pictures we were asked to select, thirty-four (34) were pictures of white children; only six (6) were pictures of black children. This division caused discussions and conclusions in our group. From the six (6) black children in the group, only two (2) of them agreed they were really black. There were thirty-one (31) white children in positive situations and three (3) in negative situations. In the group of pictures of black children (six), four (4) of them were in negative situations.

3. *Media and Racism*

As we can observe, there is a certain difficulty in defining what it is to be black or even who is black in Brazil. Voting was the chosen way for the researchers of the subgroup to overcome this "difficulty".

Over the course of several years, we termed the percentage of black or white people in publicity articles in the researched press media the "visibility degree".

The visibility degree is a ratio indicated by (), where i can vary according to the analysis interest.[15] In the landscape entitled "Media and Racism", where it was relevant to consider the visibility of black and white people, i assumed the attributes W or B, according to the way the character was classified by the students. We get:

$$VDi = \frac{number\ of\ pictures\ of\ ethnic\ group\ (i)}{number\ of\ pictures\ of\ all\ ethnic\ groups\ that\ were\ found}$$

Tables 2 and 3 summarise data from a sample of editions of this landscape and allow us to conclude that only a small amount of black people participate in advertising articles in the surveyed magazines.

Table 2. About the visibility degree. Made by the author.

YEAR	2009	2010	2013
WHITE CHILDREN	626 (85%)	538 (84%)	389 (88%)
BLACK CHILDREN	115 (15%)	104 (16%)	51 (12%)
TOTAL	741	642	440

What we termed "visibility degree" identifies the degree (percentage) of participation of the children (or characters) of each "type" in the pictures collected by the students. For example, Table 2 shows that, in 2010, out of the 642 pictures collected, 538 (84%) portrayed white children. This percentage represents the visibility of white characters in the collection made by the students.

---

15 In a discussion about gender (i) you can assume the attributes (M) male or (F) female or any other attributes that you want to observe and analyse in terms of its visibility in these printed mediums.

Table 3. The visibility of white characters in the collection made by the students. Table constructed by the author.

	2009		2010		2013	
	WHITE	BLACK	WHITE	BLACK	WHITE	BLACK
POSITIVE	91%	65%	86%	55%	88%	72%
NEGATIVE	9%	35%	14%	45%	12%	18%

Through Table 3, which refers to the assessment of the characters' contexts, it is possible to see how white people and black people are portrayed in positive or negative contexts.

In 2009, out of the 626 pictures of white children collected, 9% were described negatively while, in the same year, out of the 115 pictures of black children collected, 35% were described negatively.

Furthermore, in the results expressed in Table 2, we observe that across the years, the percentage of positive participation of white children is shown inertially inside the interval (85%—95%), while the percentage of the same nature concerning black children is inside the interval (50%—75%).

If we consider only the negative contexts of participation, white children appear in the years 2009, 2010 and 2013 respectively, at rates of 9%, 14% and 12%, while black children appearin higher percentages during the same period: 35%, 45% and 18% respectively. Unmistakably, black children are, proportionally, portrayed more often in negative situations.

In 2017, the results (Table 4) show a picture that is very similar to the previous ones, with the exception of "participation quality" (PQ) of black characters reaching 89% positive participation. From the total of 18 pictures that portrayed black characters,16 were represented in positive contexts, as described by the researchers. However, the visibility degree of black people continues to be inertially low. From the total number of pictures collected (73), only 18 represented black characters, which results in $VD_B = \frac{18}{73}$, a visibility of about 25%.

Table 4. Compiled results from the research subgroups of the SRG in 2017: Groups from ninth grade.

2017	WHITE CHARACTERS			BLACK CHARACTERS			
	PICS	POSIT.	NEGAT.	PICS	POSIT.	NEGAT.	TOTAL
G1/9B	16	15	01	03	03	00	19
G2/9B	10	09	01	01	01	00	11
G3/9B	07	07	00	03	03	00	10
G4/9B	15	13	02	03	02	01	18
G5/9B	07	06	01	08	07	01	15
SUM	55	50	05	18	16	02	73

Table 5 shows the data from 2018, compiling information from five groups of students.[16]

Table 5. Compiled results from the research subgroups of the SRG in 2018.

	TOTAL	WCP	WCP+	WCP-	BCP	BCP+	BCP-
G1	20	19	18	01	01	01	00
G2	15	15	13	02	00	00	00
G3	30	27	21	06	03	03	00
G4	24	19	16	03	05	03	02
G5	32	32	31	01	00	00	00
TOTAL	121	112	99	13	09	07	02

Let us consider what we call Mathematical Measures of Democracy (MMD)[17]—the mathematical calculation that seeks to express whether a particular practice assumes democratic aspects or not. We considered the MMD to be "thinking abstracts". These measures are "theoretical

---

16 In this table, WCP refers to White Characters' Pictures and BCP to Black Characters' Pictures.
17 The MMD initialism was a contribution from the researcher (student of the ninth grade) Giovani, alluding to the LCM (Least Common Multiple).

constructions [...] used to ease the thinking process and can be exemplified in mathematical concepts and mathematical models" (Skovsmose, 2001, p. 81). I understand the "thinking abstracts" to be theoretical constructions that were initially proposed to help us reflect and act on a certain sphere of life, but which can end up becoming "concrete abstractions", part of our routine actions.

Returning to the landscape description course, we note that in Table 5, we have:

$$VDw = \frac{112}{121} = 0.92$$

This means that 92% of the collected pictures were characters classified as white by students—a very high degree of visibility for this ethnic group. Therefore, in only 8% of the pictures collected are black characters represented.

As for the quality of participation, another mathematical measure of democracy, we have:

$$QP_i = \frac{number\ of\ pictures\ of\ characters\ (i)\ with\ positive\ surroundings}{number\ of\ pictures\ with\ characters\ from\ ethnic\ group\ (i)}$$

$$QN_i = \frac{number\ of\ pictures\ of\ characters\ (i)\ negative\ surroundings}{number\ of\ pictures\ with\ characters\ from\ ethnic\ group\ (i)}$$

Recall that i assumed, in this scenario, the attributes White (W) or Black (B). This way, for $QPw = \frac{99}{112} = 0.88$. In 88% of their pictures, white characters were represented in positive contexts (or "surroundings").

On the other hand, for $QP_B = \frac{7}{9} = 0.77$. In 77% of their pictures, black characters were described in positive contexts (or "surroundings").

4. Final Considerations

With respect to school research groups (SRGs), we have to consider the potential for producing a positive impact on the teaching and learning of mathematics processes in schools. Furthermore, SRGs have revealed themselves as important strategies for approaching racism in the mathematics classroom, and that can extend to other social themes relevant for democracy.

Printed media, spotlighted in this text and in the landscapes that were built around the investigation, contributes deeply to documenting the invisibility of black people and, thus, to revealing the corrosion of democracy, from a racial point of view. This is the very reason why we describe the production of this invisibility as an erosion of democracy.

By proposing the Mathematical Democracy Measures (MMD), we want to show the democratic status that mathematical calculations, algorithms and processes can take on the structuring and reading of various daily social practices. The degree of visibility and the quality of participation, in turn, are the calculations that allow us to reveal the invisibility of black people and, therefore, are related to racial democracy. These are the mathematical measures of racial democracy.

Finally, these SRG actions in the media and racism landscape are part of a set of studies and theoretical perspectives that have fuelled my professional practice, which is related to a growing concern for the rescue of the African and Afro-Brazilian people's contributions to the production of mathematical knowledge. For these reasons, I have named this practice *a decolonial look at the mathematics classroom*.

References

Britto, R. R. (2013). Educação matemática e democracia: Mídia e racismo. *Actas del VII CIBEM ISSN, 2301*(0797), p. 3355. http://www.cibem7.semur.edu.uy/7/actas/pdfs/568.pdf

Britto, R. R., Moura, A. Q., Nascimento, C. A. F., Roncato, C., Biotto Filho, D. & Figueiredo, M. O. R. (2017). Cenários para investigações nas salas de aulas de matemática de escolas brasileiras. *Revista paranaense de educação matemática*, 6(12). http://www.fecilcam.br/revista/index.php/rpem/article/view/1601

D'Ambrosio, U. (2001). *Etnomatemática: Elo entre as tradições e a modernidade*. Autêntica.

Domingues, P. J. (2004). *Uma história não contada: Negro, racismo e branqueamento em São Paulo no pós-abolição*. Senac.

Freire, P. (2000). *Pedagogy of the oppressed*. (Translation by M. Bergman Ramos). Continuum.

Hofbauer, A. (2006). *Uma história do branqueamento ou o negro em questão*. Unesp.

Powell, A. B. & Frankenstein, M. (Eds) (1997). *Ethnomathematics: Challenging Eurocentrism in mathematics education*. SUNY Press.

Seyferth, G. (1995). A invenção da raça e o poder discricionário dos estereótipos. *Anuário antropológico*, *18*(1), 175–203.

Skidmore, T. E. (2012). *Preto no branco: Raça e nacionalidade no pensamento brasileiro (1870–1930)*. Companhia das Letras.

Skovsmose, O. (2000). Cenários para investigação. *Bolema*, *14*, pp. 66–91 https://www.periodicos.rc.biblioteca.unesp.br/index.php/bolema/article/view/10635

Skovsmose, O. (2001). *Educação matemática crítica: A questão da democracia*. Papirus.

Skovsmose, O. (2008). *Desafios da reflexão em educação matemática crítica*. Papirus.

Van Dijk, T. A. (2012). *Discurso e contexto: Uma abordagem sócio cognitiva*. Contexto.

4. Bringing the Debate over Marijuana Legalisation into the Mathematics Classroom

Agustín Méndez and
Mario Sánchez Aguilar

> We describe a teaching experience that is based on the ongoing discussion about the legalisation of marijuana, which continues to be a hot topic in Mexico and Latin America. The teaching experience took place in a private teaching institution attended by upper-middle-class students, mainly from families with a conservative political inclination. The analysis of the teaching experience focusses on the types of reflections that were triggered among the students during its implementation, and the role of mathematics in such reflections. The results of the analysis show that some of the discussions that were triggered by this activity went beyond the walls of the classroom, involving students from other school grades or even their families. We think that this contribution may be of interest to mathematics teachers interested in practical activities inspired by the critical mathematics education perspective—particularly those that have been tried out in actual mathematics classrooms.

In this chapter we describe the implementation of a teaching activity for the mathematics classroom inspired by the perspective of critical mathematics education (Skovsmose, 2014). Our intention when designing and implementing this activity was for students to look at problems in their social environment through mathematics. Thus, we decided that the focus of the activity should be a socially relevant discussion, and therefore we selected the issue of the legalisation of marijuana, which continues to be a hot topic in Mexico.

The teaching activity that we report on is far from being a traditional activity for the mathematics classroom. In addition to being a means to analyse a socially relevant issue, the activity supports students in performing inquiries and investigative work. In this sense, the activity that we present could be considered a *landscape of investigation* (Skovsmose, 2001).

A distinctive feature of the educational experience that we report here is that the teaching activity was implemented in a private educational institution attended by upper-middle-class students. Different conditions had to be in place in order to introduce these students—who were in comfortable positions in life—to a landscape of investigation where some of their ethical values were involved and challenged.

We present an analysis of the teaching activity, focussed on the types of reflections that were elicited among the students during its implementation, and the role of mathematics in such reflections. The results show that some of the discussions that were triggered went beyond the walls of the classroom, involving students from other grades and even students' families.

Before describing the teaching activity and its implementation, next we briefly review some of the notions from the critical mathematics education approach that served as a source of inspiration and theoretical support for the development of the activity. In particular, we refer to some concerns of this approach.

1. Some Theoretical Notions to Frame Our Work

The critical mathematics education (CME) approach is a theoretical approach within the field of mathematics education research that can be described in terms of its "concerns" or issues of interest. One of its main concerns is to provide students with a mathematical education that allows them to identify, judge and criticise the uses—and misuses—of mathematics in their own social settings. Under this perspective it is intended that students use mathematics as a tool that enables them to analyse and criticise their own social reality. In particular, Skovsmose and Nielsen (2014, p. 1257) affirm that two of the concerns of CME are:

(a) Citizenship identifies schooling as including the preparation of students for being an active part of political life.

(b) Mathematics may serve as a tool for identifying and analysing critical features of society, which may be global as well as having to do with the local environment of students.

The teaching activity presented in this chapter embraces these two concerns. It is an activity through which students identify and analyse an issue that is socially relevant in their local environment, but also intersects with certain political interests and ethical values of the society to which they belong. An explicit and open discussion about these political interests and ethical values is a formative experience for students that introduces them to different worldviews and initiates them into the political life of their society.

As mentioned, the teaching activity presented in this chapter can be considered as a *landscape of investigation* since it allows students to abandon—at least temporarily—the exercise paradigm that usually predominates in their mathematics lessons. It introduces them to a scenario in which independent investigative work is promoted, as well as the elaboration of conclusions and the subsequent collective discussions.

According to Skovsmose (2001), there are at least three types of landscapes of investigation, which are distinguished from each other by the references they use to provide meaning to the activities they contain: (1) landscapes with reference to pure mathematics, (2) landscapes with reference to a semi-reality, and (3) landscapes with real-life reference. The teaching activity presented in this chapter is a landscape of investigation with real-life reference. Through this milieu of learning, students are introduced to the discussion of the legalisation of the recreational use of marijuana in Mexico. This is achieved by promoting among the students the inquiry, analysis and discussion of figures and official data on the production, distribution and consumption of marijuana in Mexico.

Another peculiarity of this landscape of investigation is the school context in which it was implemented: a private educational institution attended by upper-middle-class students. Skovsmose (2016) mentions that, although this kind of student may not be subject to social risks—and may even directly benefit from social injustices and economic inequalities—a critical approach to mathematics instruction can help them develop a critical consciousness that supports them in deepening their knowledge and understanding of the socio-political contexts of their lives (p. 3). In this chapter, we empirically explore the gains that students in comfortable positions can obtain by engaging in a landscape of investigation within their mathematics lessons.

Another key notion in this work is that of *reflection*. Several scholars argue that reflection plays a fundamental role in the mathematical development of individuals. For example, researchers in line with the Piagetian tradition (e.g. APOS Theory, Arnon et al., 2014) underline the importance of reflecting on the actions that one applies to mental (mathematical) objects, to internalise them and thus achieve more robust levels of understanding. Reflection is also considered a crucial element in the professional development of mathematics teachers (Chapman, 2008).

CME conceptualises reflections as part of a collective endeavour embedded in dialogue and interaction. According to Alrø and Skovsmose (2002), reflection means considering at a conscious level one's thoughts, feelings and actions (p. 184). Furthermore, "reflections become an essential part of a developed literacy and it takes on a political dimension by addressing the broader context of the particular elements of learning" (p. 165). This broader context of the particular elements of learning is captured through the dimensions *scope of reflections, subject of reflections* and *context of reflections*.

Of special interest for our analysis is the *scope of reflections*, which refers to what is addressed in a reflection. For instance, when a group of students is collectively solving a mathematical task, they can reflect on the reliability of their calculations, on the usefulness of solving the task, or on whether they will have to solve a similar task in the next examination. In this study, we identify and categorise the scope of reflections that emerges when students are involved in a landscape of investigation focussed on analysing and discussing the legalisation of marijuana in Mexico. We were interested in discerning the extent of the reflections elicited by this landscape of investigation.

2. Design and Implementation of the Landscape of Investigation

An activity can only function as a landscape of investigation if the students accept the invitation for it to do so (Skovsmose, 2001, p. 125). "Accepting the invitation" means that students decide to get involved in the activity, because they find it interesting, attractive or somewhat relevant. We as teachers decided that the legalisation of marijuana in Mexico would be the focus of the activity. One of the reasons for placing the issue of marijuana legalisation at the heart of the activity was to encourage a larger number of students to accept the invitation. Marijuana legalisation is a controversial topic in Mexico that can be looked at from different perspectives: economic, ethical, medical, political, social, recreational, cultural, etc. We expected that this multifaceted and socially relevant problem would arouse the interest of the students, especially in the context of the mathematics classroom.

In order to achieve some distance from the exercise paradigm, the landscape of investigation was designed by the authors of this chapter (the first author was the class teacher) as an open, collaborative, investigation-oriented activity, which was implemented over several sessions. The activity was introduced to the students through a printed document that refers to the national debate on the use of marijuana, which took place in Mexico at the beginning of 2016. The document also refers to the bill by the former Mexican president, which was sent in April 2016 to the Senate of the Republic in order to allow the medicinal and scientific use of marijuana, as well as to permit consumers to carry up to 28 grams of the drug.

The activity is based on a role-play scenario where the students serve as "advisers to the Senate of the Republic" and provide a justified answer to the following central question: "Should marijuana be legalised in Mexico?" Additionally, students had to answer supplementary questions such as: "Economically speaking, what institutions and sectors of society would be affected by the legalisation of marijuana?" and "Does marijuana consumption depend on social class?" The role of the supplementary questions was to encourage the students to explore diverse angles of the legalisation issue, sometimes with the help of mathematics.

To provide answers to the posed questions, the students had to carry out an investigation and present a report. During the introduction of the activity, students were given some web addresses where they could find figures and data related to the national debate on the use of marijuana (for instance, http://www.gob.mx/debatemarihuana); however, they were free to draw on any other source of information in order to formulate their answers. At the end of the activity, the students had to submit their answers in the form of a written report to the Senate of the Republic. They were advised to include figures, tables and graphs as part of their report. Along with the report, students had to present their answers to their classmates and teacher. A more detailed account of the design of this landscape of investigation can be found in Méndez (2018).

The landscape of investigation was implemented between April and May 2017 in a private educational institution attended by upper-middle-class students. In particular, the activity was applied in a lower-secondary-school group consisting of thirty-two students: fifteen girls and seventeen boys, with ages ranging between fourteen and fifteen years old.

The landscape of investigation was developed as part of the regular mathematics lessons—ten sessions of fifty minutes each—although several students did investigative work outside of class time. Half of the sessions took place in the school's computer lab, with the aim of facilitating the search for information on the internet; the rest of the sessions took place in the mathematics classroom, though the students were free to use their mobile devices—such as smartphones and tablets—to support the inquiry process. During the first session, the activity was explained, and the students were provided with a printed copy of its guidelines.

The students were organised into eight teams of four members each. The idea was to integrate heterogeneous working groups based on their mathematics grades during the school year; that is, we sought to have teams composed of one high-achieving student, one low-achieving student, and two students with average levels of achievement. Students were familiar with this form of collective work, since it was a common way to form teams in their school.

It is important to note that, in order to achieve the implementation of this landscape of investigation, institutional and parental support were both essential. The school principal had a supportive attitude towards the project; however, because the legalisation of marijuana is a sensitive issue for some students and their families, the principal requested written authorisation from the students' parents to allow their children to participate in the project, and to allow us to record their activity for research purposes.

3. Analysis of the Implementation

To infer the students' reflections that arose during the implementation and development of the landscape of investigation, empirical data from different sources were collected and analysed.

At the beginning of each session, one student from each work team was designated to audio-record the interactions and exchanges between the team members with a smartphone. At the end of the session, the students sent their recordings to the mathematics teacher via email, Google Classroom or WhatsApp.

In addition, the teacher who implemented the activity (the first author of this chapter) took field notes during all sessions. In these notes, certain reactions, comments, and questions that students asked during the sessions were recorded. The final presentations made by the students were also audio-recorded and their final written reports were collected for analysis.

Although the first author of this chapter led the implementation of the activity in the classroom, both authors were involved in the data analysis. The audio recordings underwent a tape-based analysis (Onwuegbuzie et al., 2009), in which we first became familiar with the data by listening to the audio recordings independently and repeatedly. Then, we discussed and identified the segments of the audio recordings where the students had reflected on the activity, and transcribed them in order to classify their scope and use the transcription as empirical evidence. It is worth mentioning that we only analysed the data from three randomly selected groups. Due to time constraints—this educational experience was developed within the time frame of a Master's degree—it was not possible to analyse the data generated by all eight groups.

4. What Kind of Reflections Emerged?

Our analysis of the data led to the identification of three categories of student reflections, each with a different scope: (1) ethical and moral reflections, (2) economic and political reflections, and (3) mathematical and technical reflections. As will be seen, these categories of reflections are not mutually exclusive, thus it is possible to identify intersections between some of them.

Next, we illustrate the students' reflections with transcripts taken from the activity dialogues and the final presentations. The students' names have been anonymised, and the transcripts translated from Spanish into English.

4.1 Ethical and Moral Reflections

These types of reflections emerged when students expressed certain values and norms regarding the consumption of marijuana. Some of these values could be considered "conservative", while other students provided alternative points of view that tended to challenge those positions or exemplify different possible scenarios:

> *Student 1*: It is very easy to influence friends and family to use marijuana, especially adolescents [...] because teenagers always want to be just like their friends, and that is a very bad thing, since marijuana is what opens the door to more addictive and harmful drugs.
>
> *Student 2*: Because marijuana itself is not so bad; what happens is that while you are stoned by the effect of marijuana, you might encourage yourself to try other types of drugs, which can be even more harmful.
>
> *Student 3*: But one of the diseases that is controlled with marijuana is Parkinson's. Scientists at the University of São Paulo detected that the cannabis-derived substance cannabidiol could lessen the associated psychosis suffered by Parkinson's patients, such as reduction of tremors, anxiety, sleep disorders and depression.

4.2 Economic and Political Reflections

These reflections arose when students made reference to economic and political aspects associated with the legalisation of marijuana that affect local populations, the state, or the country in general. An example was provided by a student, who pointed out that the country's economy would be very different if the money generated by drug trafficking was collected in the form of taxes:

> *Student 4:* I believe that the economy of our country would be very different because the consumption of drugs and all the drug trafficking generates three hundred and twenty billion dollars per year [...] most of that comes from the consumption of marijuana. If that [money] instead of going to the drug traffickers was used as taxes, those taxes would be generated by the country and that would go into the GDP, so the country's gross domestic product would be better and it would benefit many people.

Another example is the students who addressed the issue of civil liberties:

> *Student 4:* We should conclude that we must legalise marijuana because the state cannot deprive a country of its rights and freedoms. For this reason, [the state] cannot play the role of a father or try to say what is good or bad for the people [...] it is the people themselves who must know how far to go.

> *Student 3:* It has become viral or popular to say that marijuana has many benefits in comparison to its negative effects [...] it is not very important to think about its pros or cons, what we must really think about is the role that we play as a society in exercising our rights and freedoms.

4.3 Mathematical and Technical Reflections

This kind of reflection was presented mainly at the end of the teaching activity. Although the students were advised to read articles in which data, numerical tables, graphs, and percentages were presented, only a few arguments emerged in which mathematics was used. Upon being explicitly questioned during their final presentation, two working teams declared that mathematical knowledge was not very necessary for travelling through this landscape of investigation.

> *Student 4*: In this laborious investigation we have used mathematics in things such as, for example, statistics on how much money would be earned or lost with the legalisation of marijuana, and the surveys that are applied to find out the opinions of citizens about the subject.

Student 5: Well, the mathematical elements, I think they help us to know that, for example, drug trafficking generates three hundred and twenty billion dollars a year from different drugs and a quarter of that is thanks to marijuana. Then if that money did not belong to drug trafficking but was generated for the country, it would be a very different thing. Then the mathematical elements helped us to create an opinion that we will express later.

But after their inquiries and discussions, what were the students' responses to the main question: should marijuana be legalised in Mexico?

Most of the students thought that marijuana should be legalised; however, they also expressed some reservations. During their final presentations, the students highlighted advantages such as the medical potential of the herb, the economic benefits that legalisation would generate for the country, the economic weakening that organised crime would suffer, and the protection of consumers. However, the students also warned of dangers such as an eventual increase in drug use among young people and children:

Team 4: The benefits would be for people with diseases like Parkinson's. But this could affect the family environment, since children could learn to smoke, produce and even sell the product.

5. Gains and Lessons Learned: A Conclusion

The students who participated in this landscape of investigation had the opportunity to discuss and learn about a topic that is relevant to the society in which they live, and is significant for their future lives. In addition, by moving away from the exercise paradigm, the class was able to establish a forum where they listened to the opinions and positions of their peers on the subject, but also formed and expressed their own opinions based on the information examined and the collective interactions.

These students also expressed an interest—in some cases quite an unusual interest—in the mathematics lessons. The majority of the students engaged with the activity, carrying out a deep investigation of the topic, and even involving students in more advanced classes and their own relatives in their inquiries. It is fair to assume that some students considered this type of approach to be interesting and engaging, thus fostering more positive attitudes towards the mathematics class.

Before implementing the teaching activity, we had reservations about the viability of its implementation in a real mathematics classroom. That is, we thought it would be very difficult to implement such an extensive landscape of investigation (ten sessions of fifty minutes each) during regular classes in a private school. Nevertheless, one of the lessons learned was that the implementation of this type of educational approach is possible, but that strong institutional and parental support is essential.

On a more critical note, we want to point out that it was not trivial to bring mathematical content into the discussions, arguments, and reflections of the students. Although we chose a topic of study that is surrounded by information expressed in different mathematical registers (percentages, graphs, numerical tables), students rarely used these resources in their arguments and reflections. In cases where students did make use of mathematics, it was limited to concepts such as arithmetic and statistics. This trend has been reported in similar teaching experiences (e.g. Frankenstein, 2010; Reaño, 2010), thus we think that it is an aspect that should be analysed in greater detail in future experiments framed in CME.

References

Alrø, H. & Skovsmose, O. (2002). *Dialogue and learning in mathematics education. Intention, reflection, critique.* Kluwer. http://doi.org/10.1007/0-306-48016-6

Arnon, I., Cottrill, J., Dubinsky, E., Oktaç, A., Roa Fuentes, S., Trigueros, M. & Weller, K. (2014). *APOS theory. A framework for research and curriculum development in mathematics education.* Springer. http://doi.org/10.1007/978-1-4614-7966-6

Chapman, O. (2008). Imagination as a tool in mathematics teacher education. *Journal of Mathematics Teacher Education, 11*(2), 83–88. http://doi.org/10.1007/s10857-008-9074-z

Frankenstein, M. (2010). Developing a critical mathematical numeracy through real-life word problems. In U. Gellert, E. Jablonka & C. Morgan (Eds), *Proceedings of the sixth international mathematics education and society conference* (pp. 258–267). Freie Universität Berlin.

Méndez, A. (2018). *Una aplicación de la educación matemática crítica para la clase de matemáticas*. Master's thesis. Instituto Politécnico Nacional de México.

Onwuegbuzie, A. J., Dickinson, W. B., Leech, N. L. & Zoran, A. G. (2009). A qualitative framework for collecting and analyzing data in focus group research. *International Journal of Qualitative Methods, 8*(3), 1–21. http://doi.org/10.1177/160940690900800301

Reaño, N. (2010). Racist beauty canon, natural beauty and critical mathematical education. In U. Gellert, E. Jablonka & C. Morgan (Eds), *Proceedings of the sixth international mathematics education and society conference* (pp. 130–133). Freie Universität Berlin.

Skovsmose, O. (2001). Landscapes of investigation. *ZDM—Mathematics Education, 33*(4), 123–132. http://doi.org/10.1007/BF02652747

Skovsmose, O. (2014). Critical mathematics education. In S. Lerman (Ed.), *Encyclopedia of mathematics education* (pp. 116–120). Springer. http://doi.org/10.1007/978-94-007-4978-8_34

Skovsmose, O. (2016). What could critical mathematics education mean for different groups of students? *For the Learning of Mathematics, 36*(1), 2–7.

Skovsmose, O., & Nielsen, L. (2014). Critical mathematics education. In A. J. Bishop, K. Clements, C. Keitel, J. Kilpatrick & C. Laborde (Eds), *International handbook of mathematics education* (pp. 1257–1288). Springer. http://doi.org/10.1007/978-94-009-1465-0_36

5. Mathematics Embedded in Community-Based Practices: Landscape of Investigation for Examining Social (In)Justice?

Arindam Bose

> Low-income settlements with diverse income-generating work practices dispersed among households create opportunities and affordances for learning, which have mathematical elements embedded in them. Drawing from the community's rich knowledge resource (funds of knowledge), children from the neighbourhood learn to assimilate knowledge, competence, and skills. Out-of-school mathematical knowledge equips them to handle complex situations calling for quick decision-making and optimising resources and profits. Deals are negotiated following several variables, and cases of unjust wages and payments are aplenty. This chapter argues that such backdrops could form landscapes of investigation, and that the handling of or limited access to diverse goods—as well as the optimising of resources and decision-making processes—create different possibilities for not just dealing with mathematical knowledge but also questioning unfair work deals and examining social injustice. Such landscapes of investigation provide direction beyond the exercise paradigm, towards an enabling pattern of critical communication.

Mathematics is often equated with precision and rigour manifested in symbols and proofs. As a school subject, mathematics is also seen as a tool for moving towards abstraction and generalisation. Such is our belief about its esoteric position in human society that its origin in human ideas and human endeavours appears obscure and counter-intuitive. In its endeavour to be abstract and generalisable, school mathematics becomes detached from routine everyday settings and requirements. At the same time, knowledge of mathematics is valorised in the society, and is seen as an integral part of the school curriculum for all children across most cultures. However, we may argue that it is this picture of mathematics as an esoteric collection of rules and formulae, unconnected with reality, that leads it to being perceived as a subject which causes fear and a sense of failure (see National Council for Educational Research and Training, India [NCERT], 2006; NCTM Standards, 2000). It is ironic that many who study mathematics formally in schools find it difficult and frightening, and also opt out midway, but those who are outside of schools (or have dropped out) continue to successfully use mathematics in different everyday settings and in work contexts, with or without being aware of it. On the other hand, it is also well-acknowledged now that learners' socio-economic conditions influence their mathematics learning (see Bose and Kantha, 2014; Valero and Graven, 2014). There was a special issue of *ZDM—Mathematics Education* in 2014 (Issue 7) on this topic. This Special Issue and also other work in the past has explicated how children living in disadvantaged conditions often experience difficulties in learning mathematics formally in schools, and face failure which leads to dropout in many cases (also see Sarama and Clements, 2009). Incidentally, children who drop out of formal studies, as well as children from economically poor families, often get into income-generating practices. Especially in the context of the developing world, these children get access to the community's *funds of knowledge* (described in a later section), where they learn to acquire and successfully use their informal knowledge of mathematics. Economically active surroundings offer children diverse work contexts and community-based social networks, which embed in them opportunities to acquire and learn mathematical knowledge.

This chapter discusses different examples of income-generating work practice as examples of landscapes of investigation, and exemplifies the out-of-school mathematical knowledge of children immersed in such work contexts as possible tools for manifesting and investigating societal issues. In particular, I argue here that awareness of embedded mathematical knowledge in work contexts not only creates opportunities and affordances for furthering mathematics learning, but in so doing, this awareness has the potential of enabling the doers to see the "invisible", leading to the possibility of questioning the complex societal order of justice, welfare and access. This possibility of questioning plays a dual relationship of cause and effect with doers' (read "learners'") foregrounds (see Skovsmose, 2014, 2012). I show here that the contours of the landscape of investigation are drawn by the following features of the work contexts: handling diversity of goods and diverse mathematical forms embedded in them; making decisions in relation to work; optimising resources and earnings; degree of involvement in the work; awareness of linkages within and beyond the work; and associated ownership. It is important to note here that the notion of childhood and its sociology differs from culture to culture and place to place (discussed in the following section), which allows the landscape of investigation to function as a tool for existence and survival too.

Coming from an exercise paradigm (see Skovsmose, 2001; Introduction of this book), researchers have highlighted how informal knowledge gained in out-of-school contexts remains poorly built upon in classrooms—an aspect which is also attributed as a reason for the knowledge gaps among children coming from such disadvantaged conditions (See Saxe, Guberman, Gearhart, Gelman, Massey and Rogoff, 1987). These knowledge gaps are seen as "detrimental to learners' progress" during formal mathematics learning. Considering this paradigm, I argue that such informal knowledge of work-context mathematics provides children with an opportunity for exploration and explanation within and across the work domains, and for seeing through the odds of the real-life contexts that they live in. Such opportunities not only help students create meanings within activities with reference to embedded mathematics, but also provide them with avenues for enacting their identities as knowers (having specialised knowledge and skills of work details, possessed as funds of knowledge, and values attached which they eagerly share) and doers (having productive capacities that are actualised in a work setting through the products/merchandise that they make), and not just as learners (having the disposition of typical students as receivers of knowledge, as in school contexts) (see Bose, 2015). The exploration and explanation paradigm of such a landscape of investigation may help in going beyond the exercise paradigm. My hypothesis is that this paradigm can help to examine the prevailing social injustice, and show that division in real life may not always be equal, and sharing may not always be fair. The ability to realise such odds and to raise a voice against them using mathematical learning is the aim of this landscape of investigation, using out-of-school mathematics embedded in work practices. This is much on the same lines as Gandhi's vision about education centred around work (Gandhi, 1927), with concerted efforts to implement education practice that draws on knowledge related to manual vocation and work, and is draped with critical analytic skills.

A low-income settlement set in an urban, developing-world context often has its school-going students directly participate in income-generating work contexts, or is closely aware of the surrounding work contexts and practices. Experience and knowledge drawn from such contexts is intimately familiar and includes elements of mathematical knowledge, with aspects of it even embodied in students and present in the classroom. I have drawn data from my large study with Grade 6 students from two different government-run schools, with different language mediums of instruction. The schools are co-located in one school building inside a low-income settlement. Such diversity of experience within a school community presents a potentially rich opportunity for learning that has been largely ignored in formal school education.

1. Different Landscapes: Location and Work Contexts

The study was located in a large, densely populated low-income settlement in central Mumbai, with high economic viability but economically poor residents. This economically active low-income neighbourhood is spread over a two-square-kilometre area beside locations that fetch some of the highest property values (real estate) in the world (Campana, 2013). The population of the settlement is estimated to be around one million, which indicates the high population density of the locality (Campana, 2013). As a characteristic feature, the settlement has a vibrant economy in the form of micro and small enterprises dispersed among households, which include manufacturing, trade, and service units with high economic output. The entire neighbourhood generates huge employment opportunities, and being an old and established settlement, this low-income neighbourhood attracts skilled and unskilled workers from all parts of India, who come to the financial hub of Mumbai in search of livelihood. The unskilled immigrant workers find jobs in the workshops and some of them become apprentices in the small factories. Generally, the single-room, small and low-height dwellings are used for dual purposes—as workshops and as living spaces for the family and the workers. The settlement is thus a co-location of workplace and home for most of its residents. Practically every household here is involved in income-generating work and children start taking part in it from an early age. The settlement is multilingual and people from different ethnic, caste and language groups such as Hindi, Urdu, Bhojpuri, Gujarati, Marathi, Konkani, Tamil, and Telugu coexist here.

The researcher (the author) observed that almost every child in the settlement is involved in household-based economic activities, as well as in micro enterprise in the neighbourhood, in a variety of ways. Common household occupations include embroidery, zari (needle work with sequins), garment stitching, making plastic bags, leather goods (bags, wallets, purses, files, folders, belts, briefcases, waist and hand pouches, shoes), textile printing (dyeing), recycling work, pottery making, and so on. The goods produced in this settlement are not only sold in Mumbai but also exported.

2. Sociology of Childhood and Opportunity for Mathematics Learning

There are different forms of socialisation of children into adulthood, and some of these forms, especially in low-income communities, include participation in adult activity from a young age. Education policy documents[1] in India, in fact, point to participation in work as a potential corrective to the bookish and disconnected learning that happens in most schools. I have adopted the following perspective in this study:

---

1   Education policies in India are for all children—irrespective of rich, poor, or disadvantaged families. However, if a policy sees the potential of work for education aimed only at the disadvantaged children, it may constitute an approach whereby those who offer cheap labour, obey the rules and have enough physical energy will continue to be exploited. Following Paulo Freire, it is important to ask: for whom and against whom is an action?

that the participation in work practices of children from low-income communities equips them with the knowledge and identities of capable individuals, which can play a part in classroom learning and also in responding to societal issues.[2] Out-of-school experience and knowledge includes elements of mathematics, which can inform curriculum and the designing of mathematical activities and learning in school that can trigger critical questioning. Mathematics in workplace activities has distinct features as compared to the mathematics used in the formal school contexts. School mathematics is largely textbook driven which, as Freudenthal says, comes as a packaged "ready made mathematics" (Freudenthal, 1971, p. 431). Contrastingly, however, one may draw a parallel between what Freudenthal calls "mathematics as an activity"— which is about creating and applying mathematics as part of the "local organization" (p. 431)—and mathematics embedded in work contexts, in the sense that both the "activities" include a notion of relating ideas to practice. This larger study has shown (see Bose, 2015) that although mathematics in work-context-related activities remains implicit and hidden, it is hands-on and uses construction in varied ways. Though mostly routine and fragmented in nature, some of the work contexts entail on-the-spot decision-making and optimalisation that goes beyond "ready-made mathematics" prescribed in schools and comes closer to "mathematics as an activity".

Educational thinkers in the developing world, and particularly in India, have recognised the value of work experience for education conceived in a broad sense. Policy documents on education have taken on board this insight. Educational philosophers, such as Gandhi, thought of productive work as central to education, and developed a vision of education centred around work. Gandhi emphasised that modern education centred around work is different from the traditional education in the crafts. The aim of his educational philosophy of *Basic Education* or *Nai Taleem* was not training in a particular craft, but a well-rounded education of the mind, the body and the heart (Fagg,

---

2   A common critique is that the socialisation of children from the disadvantaged communities into adult life-worlds and practices is a tool for exclusion from their right to have a childhood, compared to children from other childhoods who have favourable dispositions to engage in and put intentionality in studies and in mathematics education. In this case, children from disadvantaged communities come with work socialisations and technical knowledge which are important for connecting their socio-cultural practices with mathematics learning.

2002). Under Gandhi's *Nai Taleem or "New Education"*3 (Gandhi, 1951), education—including the learning of mathematics—was to be given through the medium of crafts, which implied productive work, since Gandhi was advocating a self-supporting system of education (NCERT, 2007). However, this must not be equated with child labour; rather, different forms of sociology of childhood[4] should be nurtured in different communities and societies.

Children living in the settlement get engaged in economic activities from an early age, not only because of economic compulsions, but also because learning hand skills and engaging in income-generating practices is seen in the community as part of childhood. It is central to children's upbringing and their learning. There are parents who do not let their children work until they finish their studies, but even such children who do not participate in work develop a fair knowledge and realistic perspective on the activities and diverse work contexts around them, by virtue of the high levels of social interaction prevalent in the neighbourhood. The researcher noted that, in many cases, families (often along with their extended members) worked as a production unit, where many family members were engaged in the household-based workshops or manufacturing units. Here, family has a wider meaning, comprising of the neighbours and elders in the community.

3. Different Landscapes of Using and Learning Mathematics

In this chapter, "landscape" is referred to as the wide canvas for acting and reflecting on using mathematics. Hence, a landscape for learning mathematics and for using it can be seen as any avenue, formal or informal, at any site of action—workplace, cultural paradigms, classroom space, playground, recreational games and puzzles—wherever mathematics emerges as a tool for action. The following sections discuss different landscapes which have mathematics embedded in them, and different studies which uncovered mathematics from those landscapes.

---

3   The idea of "New or Basic Education" was formulated under Gandhi's inspiration and guidance at an Educational Conference in Wardha, India, in 1937.

4   A broader argument is the importance of connecting children's social practices with their mathematics learning and that, in the case studied, such social practices also involve socialisation at work.

The purpose is to trace different milieus of learning, which can provide a critical dimension to the purposes of mathematics learning and impact upon formal mathematics education.

4. Mathematics in Diverse Work Contexts

Research on everyday mathematics or out-of-school mathematics or mathematics in cultural practices have all unravelled different forms of landscapes for acquiring and investigating different forms of mathematical learning (see Carraher, Carraher and Schliemann, 1985; Lave, 1977; Resnick, 1987; Scribner, 1984; Saxe, 1988, 1992). Some of these studies pioneered and also influenced the emergent field of research in cultural psychology as well as in ethnomathematics. But, more importantly, these studies introduced a new perspective on what could act as the site of mathematics learning, other than formal schools. These "alternative" sites included out-of-school contexts, community-based work contexts, cultural practices and so on. Many of these studies explicated procedures which turned out to be widely different from those adopted in the formal pedagogic practices. For example, Scribner's study (1984) showed the purposive nature of the workers' cognitive strategies—they looked for ways to accomplish a task that were flexible and which minimised their effort. In addition, the workers' actions were found to reflect less application of formalisms learnt at schools, while engaging in complex mental processes for their actions (Rogoff, 1997). Scribner illustrated that workers often sought such modes of solution that were economical, required less effort and fewer steps, and had the least complex procedures. Different sets of workers—viz., assemblers, inventory staff and delivery drivers—had different ways of accomplishing their tasks and problems, but they all indicated use of "effort-saving strategies" and "procedures for simplifying and shortening solutions in different problem settings" (Scribner, 1986, pp. 25, 26). She argued that "social knowledge is incorporated in the way dairy products, for example, are stacked in the warehouse: milk, cheese, and fruit drinks are not distributed at random" but stacked at particular locations depending upon the "proximity to where they are packed, proximity to similar items and floor space" (Scribner, 1984, p. 204). This is a practical way to find an easy way of dairy product distribution. For our purposes, these work contexts can be viewed as landscapes for investigating action strategies and modes for solving the problem at hand. As discussed above, such solution modes looked for ways to make efforts lesser and optimal and procedures that were less complex. Such modes of solution are congruent to those adopted in formal mathematics pedagogical practices and, for the purposes of exploration in this chapter, examples of landscape of investigation.

Another example of a landscape emerged from Lave's work (1988), which highlighted the prevalence of a "curriculum" that follows different stages of task learning (from novice to expert) in the traditional forms of tailoring task in Liberia. Here, a sequence of training was followed, and an evaluation by experts or senior co-workers through observation was routinely carried out, providing immediate feedback. Opportunity for self-correction was another feature. Lave saw worksite training as an alternative to school-based job training or vocational training.

Another example of landscape is Nunes' study of construction foremen's tasks, based on the proportionality of wall measurements. The blueprint of the plan did not contain the scale of measurement, unlike the routine practice, and instead contained three measures of the walls of a room. The foremen were required to find the measure for the fourth wall by finding the scale using a pair of measures. The task was just the inverse of their usual practice but the context was familiar. The foremen could reverse the operations in order to solve the task. The strategies that they used preserved the meaning of their routine tasks. However, for the control group of seventh-grade students who knew the "rule of three" for solving proportional problems, the task was not familiar, as they had not seen scale-drawings in their texts. Interestingly, students used their everyday knowledge to find solutions, and not the strategies learned at school.

Other studies have shown that, while using mathematics in everyday contexts, the doers have a continuous engagement with the objects and the situation, and they do not burden themselves with the extra effort to remember the algorithms, calculation techniques and reasoning used. In contrast, school mathematics aims at mastering computational proficiency and symbol manipulation, following the correct steps in the algorithms without much freedom to use alternate techniques (Resnick, 1987). While socially shared knowledge and situation-specific competencies are the hallmarks of everyday mathematics, school mathematics emphasises improving individuals' performance and skills. Situation-specific competencies developed in everyday contexts are often linked with "physical referents" and with the "socio-cultural meanings" that the contexts carry (Schliemann, 1998). Such competencies reflect consistent logical/mathematical principles across different work contexts.

These landscapes of using mathematics in action suggest that situational variables influence tendencies in problem-solving, and that concrete problem situations like in the foremen's problem tempt the doer to use oral computation procedures drawn from everyday knowledge, whereas routine computations encourage the use of school-taught computation algorithms. In other words, such references to real life situations provided milieus for learning, and that too from a critical dimension. As Skovsmose (2001) argues, "different milieus of learning may provide new resources for making the students both acting and reflecting" (p. 123).

All the above-mentioned studies illustrate the variety of arithmetic procedures that are part of out-of-school mathematics, and at the same time discard the idea that arithmetical competencies can only be obtained in schools (Greiffenhagen and Sharrock, 2008). In fact, one of the major contributions of the studies done on out-of-school (or everyday or street) mathematics in mathematics education research (MER) and other areas of anthropology or cognitive science has been to challenge the prevalent notion that considers "western" school as the only platform for acquiring mathematical competency. These studies have illustrated the diverse arithmetical competencies and procedures that arise in such out-of-school locations which have strong potential to inform school mathematics learning. As Lave (1988) puts it, these studies "discovered mathematics in both non-Western traditional societies as well as the academic hinterland of the West" (p. 3). Examples discussed in the later section indicate that the familiar everyday and work contexts entail potentially strong resources that can scaffold mathematics learning in schools (see Bose, 2015).

5. Distinction Framework

Table 1 summarises the distinctions between school and out-of-school mathematics that have emerged in the literature. This table also serves to present a framework to distinguish out-of-school and school mathematics in the analysis.

Table 1. Distinctions Between Out-of-School Mathematics and School Mathematics.

Difference	Out-of-School Mathematics	School Mathematics
Cognitive features	-Based on shared cognition (Resnick, 1987) -Manipulations are carried out using quantities -Use of group work and division of labour (Resnick, 1987) -Use of tool manipulations	-Based on individual cognition (Resnick, 1987) -Manipulations are carried out using symbols (Resnick, 1987) -Individual, independent work -Use of pure mentation
Intended outcomes of learning	-Situation specific competencies	-Generalised learning, power of transfer (Resnick, 1987)
Difference in numeration/ procedures	-Oral -Use of multiple units and operations (Saxe, 1988) -Use of contextualised reasoning (Resnick, 1987) -Use of decomposition and repeated groupings (Carraher et al., 1987) -Use of convenient numbers (Carraher, et al., 1985)	-Written -Use of symbols -Use of formal reasoning -Use of formal algorithms taught in schools
Mechanisms of acquiring knowledge	-Communication, sharing, legitimate peripheral participation (Lave and Wenger, 1991) -Learning from one another, circulates in communication, role of artefacts and language (Carraher et al., 1987)	-Knowledge acquisition and knowledge building are textbook-based -Based on individual thinking, group-work is not always encouraged

Difference	Out-of-School Mathematics	School Mathematics
Meta-cognitive awareness	-Meaningfulness, confidence in procedures and obtained results (Nunes et al., 1993; Saxe, 1988a) -Continuous monitoring ("where they are" in the middle of calculations) (Carraher et al., 1987)	-Heavy use of algorithms, lack of meaningfulness and relevance -Continuous monitoring usually not possible
Test of the acquired knowledge	-No formal examination -Tested by seniors/experts through observations	-Use of formal examinations, consisting of mostly written tests

The distinction framework helps in identifying instances of out-of-school mathematics in work practices, and I show below how these instances form landscapes of investigation. These instances are embedded in the cultural practices, and unpacking them requires knowledge of rich cultural resources called "funds of knowledge". The hope is to acquire an inquisitive tool in the form of landscapes of investigation for countering constant denial, prolonged oppression and unjust treatment.

6. Funds of Knowledge

It is widely seen that children in low-income conglomerations are often bound in social relationships and work practices from an early age, and the broad features of their learning develop at home as well as in their surroundings. Households and their surroundings contain resources of knowledge and cultural insights that anthropologists have termed *funds of knowledge* (Gonzalez, Moll and Amanti, 2005; Moll, Amanti, Neff and Gonzalez, 1992; Velez-Ibanez and Greenberg, 2005). The "funds of knowledge" perspective brings insights to mathematics education research that are related to, but different from, the perspectives embedded in studies of "culture and mathematics". In contrast to restrictive and sometimes reified notions of "culture", "funds of knowledge" emphasise the hybridity of cultures and the notion of "practice" as "what people do and what they say about what they do"

(Gonzalez, 2005, p. 40). The perspective also opens up possibilities of teachers drawing on such funds of knowledge and relating it to the work of the classroom (Moll et al., 1992).

Funds of knowledge are acknowledged to be broad and diverse. They are embedded in networks of relationships that are often thick and multi-stranded, in the sense that one may be related to the same person in multiple ways, and that one may interact with the same person for different kinds of knowledge. In other words, funds of knowledge point to the diversity of contexts and settings from which knowledge is acquired. Such funds are also connected and reciprocal. When they are not readily available within households, then they are drawn from outside the household, from networks in the community. The concept thus emphasises social interdependence. Furthermore, from the funds of knowledge perspective, children in households are active participants, not passive bystanders.

I have used the notion of funds of knowledge as a guiding notion in analysing the work contexts that students are exposed to, and in illuminating the nature and extent of everyday mathematical knowledge available within the community of the classroom. Funds of knowledge in this work are seen as a resource pool that emerges from people's life experiences and is available to members of the group, which could include households, communities or neighbourhoods. In a situation where people frequently change jobs and look for better wages and possibilities, members of the household need to possess a wide range of complex knowledge and skills to cope and adapt with the changing circumstances and work contexts. Such a knowledge base becomes necessary to avoid reliance and dependence on experts or specialists, particularly in jobs that require maintenance of machines and equipment.

Access to school education that is seen as meaningful and relevant by underprivileged communities and is connected to their life settings has remained elusive. What is offered in schools at present is a structured educational package detached from most students' everyday life experiences, yet accepted as "legitimate" knowledge since it acts as the "gate-keeper" (Skovsmose, 2005) to different kinds of opportunities and future social well-being. The legitimacy and necessity of the landscape of "formal" school mathematics renders all other forms of mathematical knowledge not only insignificant but also ineffective. This landscape comes as a "package" of formal school mathematics which is seen as beyond challenging or above questioning. That apart, this "package" either repels or attracts people, depending largely on their socio-economic status. In this backdrop, it is widely accepted that hierarchical social structure (for example, caste and class division in Indian society) has bearings on how this landscape is used. I discuss below examples where, through using their community's funds of knowledge, middle-graders engaged in different work practices and used mathematics as landscapes of investigation, with or without being aware of it.

7. Mathematics in Diverse Work Contexts: Landscapes of Investigation

Features of work contexts and the degree of students' engagement in them shape the learning experiences of students who participate in work contexts and the richness of the knowledge that they acquire. Landscape of investigation that I refer to emerge from the aspects of and ways in which work contexts create opportunities and affordances for (mathematics) learning and for viewing the world through a critical lens. I describe below six examples of such landscapes. One common characteristic feature of these landscapes is the diversity of goods or artefacts used in the work contexts. Some kinds of work involve diverse interactions with people and material, leading to greater opportunities to acquire knowledge and skill, while some tasks are fragmented and create limited or no scope for interaction with people. Thus one can find economic activity that is varied and calls on a range of knowledge and skills, as well as activity that is routinised, making little demand on skills and knowledge. Hence, one of the factors in creating opportunities for varied learning experiences is the nature of work itself, which may be difficult and demanding, or repetitive and mechanical, but is characterised by diversity (Bose and Subramaniam, 2013).

8. Handling Diversity of Goods and Diverse Mathematical Forms

The question of how diversity of goods is dealt with in a work context gave rise to exploration, investigation and learning opportunities. For example, Salman's (13 years old, sixth-grader) mobile repairing work calls for interactions with different people, from negotiations with a variety of customers who bring the defective mobiles in, to those from whom he buys spare parts (all names are pseudonyms). Salman has knowledge of the spares and tools market in connection to his work—with shops selling mobile phone spares, parts, repairing tools, electrical appliances like soldering machines, and other instruments. He has knowledge of different mobile parts and their functions; he knows the costs of both original spare parts and low-cost substitute parts that are made locally, and about different products of different brands and their prices. He travels to far-away markets which sell spares at lower prices than the neighbourhood shops. Knowledge of a range of products and brands and their prices is required for his work. Salman, in the process of his work, deals with a diversity of numbers—such as cost prices, number of materials and the number of each of them, etc.

The second example is of Rizwi (13 years old, male, sixth-grader) who knows about the different raw materials used in his textile printing (called "dyeing") work—viz. stoppers (blocks used in printing the design), dyeing frames of varying sizes, dyes producing glossy or matt finish of different colours, thinner, adhesive, coating material, their prices and the units in which they are sold. He knows the sizes in inches, which is an old British unit and no longer part of the school curriculum. He explained that the typical frame-sizes are 16"×12", 28"×12", which cost between Rs 2000–3000 to make. He knows which colours are to be mixed to obtain a particular colour or shade, and the proportion in which they are mixed. Rizwi informed the researcher that he makes a choice of the suitable dye-frame and stopper of a particular size by looking at the logo design to be printed in a given task. The unit used for measuring is inch. When asked to make a "guess" of the dimensions of a few objects lying around, viz. voice recorder, desk, notebooks, he gave nearly accurate answers—all in inches.

The third example is the work of fixing stones on jewellery (earrings, buckles, pendants), which is usually done at home. This also reflects diversity, although is limited in comparison with other kinds of work. Those doing this work, like Rani (12 years old, female, sixth-grader), have to deal with pieces of jewellery of different kinds, upon which a certain number of stones need to be fixed. Stone-fixing work involves putting coloured stones (usually up to four) on earrings, pendants, rings, buckles, and mangal-sutra (a kind of necklace). This is a fragmented task and Rani did not need to have knowledge about other parts of the production network (such as where the jewellery is made, where it is sold, where the raw materials are produced, and so on) in order to do it. Although she was aware of some of these linkages and mentioned them to the researcher, such knowledge was not common among other students engaged in household-based fragmented tasks. The variation in such pieces is also an occasion for learning. The finished pieces are mounted on card in different arrangements that Rani is intimately familiar with, and she associates multiplication facts with these arrangements. Rani knows about the wages for fixing different numbers of stones, different types of adhesives, the proportion in which they are mixed and their prices (Excerpt 1).

Excerpt 1. Interview transcript: Rani explaining her wage. Source: Bose, 2015.

104[5]	S	one forty four jodi banayenge na tho one gurus hota hai tho usme eighteen rupees hi milte hain/ gurus ka full packing karenge na thaile mein daal ke...	if we make one forty four pairs then it's one gurus (gross) for which we get eighteen rupees/ if we pack full gurus in a packet.
105	T	haan, haan/	yes, yes/
106	S	to one gurus matlab one packing ka to twenty four cards honge, ... matlab one card mein six jodis lagte hain, uss tarah na apun ... twenty four cards ka packing karenge na, to apne ko nineteen rupees milega aur nahin packing karenge to eighteen rupees hi milega/	then for packing one gurus, one packing has twenty four cards, ... [and] one card has six pairs, ... if we pack twenty four cards, then we get nineteen rupees and if we don't do packing we get eighteen rupees only/

Stone-fixing work carries an extra rupee for packaging. For example, for fixing three-stone, one-gross earrings, Rani gets Rs 18, and for packing

---

5   In all excerpts, "S" and "T" represent student and researcher respectively; while the numbers on the left column indicate the line numbers in the original transcript, and the English translations of the utterances are provided in the right column.

all the twelve or six cards together (totalling one gross), she earns an extra rupee, i.e., Rs 19. Though it is a meagre amount, Rani makes the effort to earn the extra rupee. The adhesive used in the stone-fixing work is purchased by the workers. Rani explained that two different kinds of adhesives are bought, yellow and white, and while pasting they are mixed together. Rani could not explain how much adhesive one bottle contained. However, she estimated that one bottle can be used to make up to 200 "gurus" of earrings.

9. Making Decisions in Relation to Work

In the mobile phone repairing work, Salman has to quote a price for a job by guessing the customer's paying capacity. This helps in deciding whether to use an original part or a low-cost substitute, keeping in mind the expected profit. Quoting a price may often call for mental computation of quantity and price of required parts, and the time required to carry out repairs. Salman also has to keep in mind what other shopkeepers in the vicinity are charging for the repair work. Thus, Salman's decisions are similar to the ones made by his friend, the repair shop owner.

In Ehsan's (13 years old, male, sixth-grader) garment recycling work, making quick decisions is of vital importance. Negotiations happen during the quotation of prices, which are competitive because of the presence of many groups of waste collectors in the field, who are also into the collection of *chindhi* (leftover cloth pieces of varying sizes collected from the garment-stitching workshops). Hence, the price negotiation with the buyer for selling the collected *chindhi* is done quickly and it is therefore imperative for Ehsan to make quick decisions. For this, he claimed he uses quick computations since his job calls for cross-checking the amount quoted by the buyer. He also explained that his personal contacts with people on the network help him in getting a good price. During negotiation, Ehsan negotiates price based on different variables such as total weight, cloth-piece colour, cloth-piece size, texture, quality (cotton, silk, synthetic, etc.), and time spent. Decisions on price quote and entire price negotiation happen through quick mental calculations, often involving complex numbers without use of any pen, paper or calculation device.

Ehsan used proportional reasoning while doing various types of calculations. He collected *chindhi* once a week and sold them at rates varying between Rs 4 and 20 a kilo (kg). White *chindhi* fetched the highest value of around Rs 20 per kilo, while red coloured ones were sold for Rs 8 a kilo. *Chindhi* of smaller dimensions fetched Rs 4 per kilo. Ehsan visited around 25–30 workshops per visit, often with a group of boys of a similar age (his friends or known to him) and collected up to 70 kilos of *chindhi* in a day (his highest collection was 95 kilos in a day). Proportional reasoning was used while distributing (or sharing) the wage among themselves. This is especially so in a situation when the *chindhi* collected by the whole group was weighed as a whole, and the amount was paid together and not individually. Ehsan explained that he calculated an individual's earning based on estimates of the amount of *chindhi* he had collected. *Chindhi* collectors like Ehsan did not have control over the kind of "maal" (material/goods) they would get on a given day, but in order to make more money while carrying a fixed amount of weight, Ehsan used proportional reasoning to make decisions about how much to collect of the white variety and the coloured variety. He was aware that collecting the white variety fetched him five times more than the ordinary or smaller pieces of *chindhi* (20:4), while the red variety brought only double the amount of the smaller *chindhi* (8:4).

10. Optimising Resources and Earnings

Ehsan's garment recycling work entails remembering many numbers and proportions of the material sold under each type while calculating the wage. This is reflected in Excerpt 2 where Ehsan spoke about an instance when he earned Rs 640 by collecting 95 kilos of *chindhi* on a single day.

Excerpt 2. Interview transcript: Ehsan explaining his work calculation.
Source: Bose, 2015.

364	S	sabse zyada to bahut pehle kamaya tha, chhe sau chalis rupaya/	maximum earning was long back, six hundred forty rupees/
365	T	kitna kilo tha usme?	how many kilos were in it?
367	S	panchanve kilo/	ninety five kilo/

368	T	panchanve kilo ka chhe sau chalis rupaya? Kaise?	six hundred forty rupees for ninety five kilo? how?
369	S	...white alag, lal alag kar liya, white becha tha bees rupaya kilo, aur lal becha tha aath rupaya kilo aur chindhi becha tha (char rupaya kilo)/	...white and red separated, sold white for twenty rupees a kilo, and sold red for eight rupees a kilo and sold chindhi (four rupees a kilo)/

Ehsan's work entailed optimisation skills on several variables including optimality of time, traverse and weight to carry is considered. *Chindhi* collectors earmarked workshops where they are most likely to get a handsome amount and know how quickly they can fix a deal before other groups can drop in. One also has to keep in mind the weight that can be carried easily and which can fetch a good amount.

11. Understanding Numbers to Understand Fairness and Social (In)Justice

I argue here that access to out-of-school mathematical knowledge and exposure to work-context knowledge provides different features of landscapes for investigating various societal issues at the intersection of one's social location and work practice. Landscapes of investigation using numbers for understanding the notion of fairness, equality and justice are a route for achieving the broader goal of formal mathematics education and for empowering mathematics learning. Examples are aplenty where such a realisation appears to be a distant dream. For example, students engaged in work practices often make judgements about the fairness of a deal. However, this study also indicated that students who are learning certain tasks (like Abedin as a novice in a tailoring workshop) or household-based "chunked" work that involved "making" (like rakhi-making, stone-fixing work, latkan-making work) or fragmented tasks of a larger chain of work (like cloth-cutting in tailoring tasks) often cannot question the wages that come their way. It seems that the easy availability of cheap labour in the low-income settlement is always more than the demand, which forces the workers to accept the wages given, however meagre they are. When Abedin (14 years old, male, sixth-grader) was asked to discuss what he felt about the amount he got as an early apprentice, he replied, "sahi hai sir" [it's correct sir]. Further probing about whether it should be more or less is quoted in Excerpt 3:

Excerpt 3. Interview excerpt: Abedin explaining his work calculations. Source: Bose, 2015.

394	S	jyada kaam rahega na sir to aisa lagta if there's more work sir then it seems hai ki paisa kum de raha hai/	if there's more work sir then it seems hai ki paisa kum de raha hai/ as if the wage is less/
395	T	achha, aisa lagta hai ki paisa kum de raha hai/	okay, you feel that you are paid less/
396	S	asteen mein jaise, pyjame mein ho gaya, asteen mein ho gaya, to lagta hai paisa kum de raha hai, do rupaya-do rupaya ka dega/	like in asteen (cuff), or in pyjama, it seems as if we're getting less, two rupees—two rupees is given/
397	T	achha../	okay.../
398	S	haan, aur sirf gale ka rahega to aisa lagta hai paisa sahi de raha hai, ek rupaya/	yes, for only neck it seems the wage is correct, one rupee/
399	T	aisa kyon? Matlab kum lagta hai, jyada lagta hai, aisa kyon?	why so? you feel less or more, why so?
402	S	jyada kaam ho jata hai to itna sara kaam dekh ke aur paisa sirf do rupaya milta hai to sahi nahin lagta hai/	when there's more work then by seeing so much of work and amount given is only two rupees then we feel it's not correct/

12. So What?

Much to the regret of educators, school mathematics curricula—which often aim at building meaning-making and mathemacy skills using real-life contexts, and which middle-graders in this study learnt in their school—actually fail to develop any critical analytical skills in mathematics among students. They could not start questioning their predicament, which is bound with gross injustice, sheer exploitation and perjury. School mathematics, instead of providing the promised

5. Mathematics Embedded in Community-Based Practices 89

so-called equitable opportunities, does not even create a safe space for meaning-making using everyday contexts of economic exploitation in terms of wage calculation, work-hour calculation and optimising resources—not even to think of profit-making! Thus school mathematics failed to provide the platform for questioning which we are referring to as a landscape of investigation in this book. Discussion on low-wage distribution with one student in the sample failed to evoke any response, possibly because of the underlying power structure. Sahana, a 13-year-old female student, did not seem to realise the injustice and the low wage given to her and to other workers like her. Instead, her response indicated *justification* for the wage structure. She claimed that other costs were incurred before the material went to the retail shopkeepers for selling. According to her, there were underlying invisible costs like, those incurred during dispatch, designing and polishing charges, and others' wages. She added that she was "happy" with the wage she was getting (see Excerpt 4).

Excerpt 4. Interview excerpt: Sahana's justification for her low wage. Source: Bose, 2015.

346	T	one twenty rupees mein becha/ to tumko sirf three rupees mila aur usne kitne mein becha?	sold for one twenty rupees/ so you got only three rupees and how much was it sold for?
347	S	one twenty/	one twenty/
348	T	one twenty mein/	for one twenty/
349	T	to yeh tumko kya lagta hai?	so what do you think about it?
350	S	unlogon ko profit nahin hota hai sir, kyunki humlog yahan pe banate hain, humlog company mein dete hain, company se tempo mein jata hai, wahan se display matlab dusre aadmi ke paas jata hai, wahan se bhav badh-badh ke wohlog ke paas aata hai, wohlog ke paas bhi...	they don't have profit sir because we work here, we hand over to the company, from company all this goes in a tempo, from there it goes for display, i.e., goes to some other person, from there the rate rises higher and higher and comes to the sellers, sellers too...

351	T	lekin phir bhi, tumhe nahin lagta ki ek sau...	even then, don't you feel one hundred...
352	S	unlogon ko jyada profit nahin rahta hai, unlog four rupees fifty paisa mein kharidte hain to five rupees mein bechte hain, unlogon ko bhi aise hee hota hai/	they don't have much profit, they buy [one pair] for four rupees fifty paisa and sell for five rupees, they too need to face this/
...	...	...	...
356	S	sir, wohlog jo company se humlog laate hain, woh bhi design banata hai, to design ko polish karta hai, polish karke design ko card mein lagane ka hota hai, ..., to unlogon ko bhi paisa dena rahta hai, to isliye bhav karke hee market mein jata hai/	sir, the company from where we bring [work], they make designs too, then they polish [refine] the design, after polishing the designs are put in the cards, ..., so they're required to be paid too, so, the rate rises and rises before it [material] goes to the market/
357	T	achha, to tumko aisa lagta hai ki yeh bhav jo tumko mil raha hai woh,	okay, so you feel that the rate that you're getting is,
358	S	Sahi hai/	it is fair/
359	T	sahi hai?	is it fair?
360	S	(nodding in affirmation)	(nodding in affirmation)
361	T	achha/ kum nahin lagta tumko?	okay/ don't you feel it's less?
362	S	(nodding in 'no')	(nodding in negation)
363	T	Sahi lagta hai? tum khush ho iss bhav se?	feel it's fair? are you happy with this rate?
364	S	(nodding in 'yes')	(nodding in affirmation)

The notions of equity and fairness, intertwined with the concepts of justice and rights, had not made a mark while she used basic arithmetical calculations. Textbook-type word problems only elicited computational skills, but did not provide opportunities to see and challenge prevailing social (in)justice. As part of oppression, the oppressed lose a sense of being oppressed. Landscapes of investigation must therefore explore how the language of education can become a language of hope that open up possibilities of social inclusion and a better life for every child (Bose and Kantha, 2014, p. 1083). Landscapes discussed in this chapter provide direction for going beyond the exercise paradigm followed in the formal learning sequence and move towards an enabling pattern of critical communication.

References

Abreu, G. de. (2008). From mathematics learning out-of-school to multicultural classrooms: A cultural psychology perspective. In L. D. English (Ed.), *Handbook of international research in mathematics education. Second edition* (pp. 352–384). Routledge.

Bose, A. (2015). *Work, knowledge and identity: Implications for school learning of out-of-school mathematical knowledge.* Unpublished doctoral dissertation. Homi Bhabha Centre for Science Education, TIFR.

Bose, A. & Kantha, V. K. (2014). Influence of socio-economic background and cultural practices on mathematics education in India: A contemporary overview in historical perspective. *ZDM—Mathematics Education, 46*(7), 1073–1084.

Bose, A. & Subramaniam, K. (2013). Characterising work-contexts from a mathematics learning perspective. In Nagarjuna, G., A. Jamakhandi & E. M. Sam (Eds). *Proceedings of epiSTEME-5: Third international conference to review research on science, technology and mathematics education* (pp. 173–179). CinnamonTeal.

Campana, J. (Ed.). (2013). *Dharavi: The city within.* Harper Collins.

Carraher, T. N., Carraher, D. W. & Schliemann, A. D. (1985). Mathematics in the streets and in schools. *British Journal of Developmental Psychology, 3*(1), 21–29.

Fagg, H. (2002). *Back to the sources: A study of Gandhi's basic education.* National Book Trust.

Freudenthal, H. (1971). Geometry between the devil and the deep sea. *Educational Studies in Mathematics, 3*, 413–435. https://doi.org/10.1007/BF00302305

Gandhi, M. K. (1951). *Basic education.* Navjivan Publications.

Gandhi, M. K. (1927). *An autobiography or the story of my experiments with truth.* Navjivan Publishing House.

Gandhi, M. K. (1951). *Basic education.* Navjivan Publications.

González, N. (2005). Beyond culture: The hybridity of funds of knowledge. In N. González, L. C. Moll & C. Amanti (Eds). *Funds of knowledge: Theorizing practices in households, communities, and classrooms* (pp. 29–46). Lawrence Erlbaum.

González, N., Moll, L. C. & Amanti, C. (Eds). (2005). *Funds of knowledge: Theorizing practices in households, communities, and classrooms.* Lawrence Erlbaum.

Lave, J. (1977). Cognitive consequences of traditional apprenticeship training in West Africa. *Anthropology & Education Quarterly, 8*(3), 177–180.

Lave, J. (1988). *Cognition in Practice: Mind, mathematics and culture in everyday life.* Cambridge University Press.

Moll, L. C., Amanti, C., Neff, D. & Gonzalez, N. (1992). Funds of knowledge for teaching: Using a qualitative approach to connect homes and classrooms. *Theory into Practice, 31*(2), 132–141.

National Council of Teachers of Mathematics. (2000). *Principles and standards for school mathematics.* NCTM.

National Council of Educational Research and Training. (2006). *Position paper of national focus group on teaching of mathematics.* NCERT.

National Council of Educational Research and Training. (2007). *Position paper of national focus group on work and education.* NCERT.

Resnick, L. B. (1987). Learning in school and out. *Educational Researcher, 16*(9), 13–20.

Rogoff, B. (1997). Foreword. In E. Tobach, R. J. Falmagne, M. B. Parlee, L. M. W. Martin & A. S. Kapelman (Eds), *Mind and social practice: Selected writings of Sylvia Scribner.* Cambridge University Press.

Sarama, J., & Clements, D. H. (2009). *Early childhood mathematics education research: Learning trajectories for young children (Studies in Mathematics Thinking and Learning).* Routledge.

Saxe, G. B., Guberman, S. R., Gearhart, M., Gelman, R., Massey, C. M. & Rogoff, B. (1987). Social processes in early number development. *Monographs of the Society for Research in Child Development, 52*(2), i-162.

Scribner, S. (1984). Studying working intelligence. In B. Rogoff & J. Lave (Eds), *Everyday cognition: Its development in social contexts* (pp. 9–40). Harvard University Press.

Scribner, S. (1986). Thinking in action: Some characteristics of practical thought. In R. J. Sternberg & R. K. Wagner (Eds), *Practical intelligence: Nature and origins of competence in the everyday world* (pp. 13–30). Cambridge University Press.

Skovsmose, O. (2001). Landscapes of investigation. *ZDM—Mathematics Education, 33*(4), 123–132.

Skovsmose, O. (2005). *Travelling through education: Uncertainty, mathematics, responsibility.* Sense Publishers.

Skovsmose, O. (2012). *An invitation to critical mathematics education*. Springer Science & Business Media.

Skovsmose, O. (2014). *Foregrounds: Opaque stories about learning*. Sense Publishers.

Skovsmose, O. O. (2021). Introduction chapter (this volume).

Valero, P. & Graven, M. (2014). Trends in researching the socioeconomic influences on mathematical achievement. *ZDM—Mathematics Education*, 46(7), 977–986.

Vélez-Ibáñez, C. & Greenberg, J. (2005). Formation and transformation of funds of knowledge. In N. González, L. C. Moll & C. Amanti, C. (Eds), *Funds of knowledge: Theorizing practices in households, communities, and classrooms* (pp. 47–70). Lawrence Erlbaum.

6. Aspects of Democracy in Different Contexts of Mathematics Classes

Raquel Milani, Ana Carolina Faustino, Lessandra Marcelly Sousa da Silva, Débora Vieira de Souza-Carneiro, Jeimy Cortés and Reginaldo Ramos de Britto

This chapter aims to discuss relations between aspects of democracy and different mathematics classrooms with the potential to become landscapes of investigation. The theoretical framework is based on the concepts of mathemacy, dialogue, representativeness, equity, inclusion, collaboration and criticism, which are essential for the maintenance and promotion of democracy. Three episodes of different mathematics classrooms are reflected upon: a class on functions in Brazilian higher education, high-school students visiting a quilombola community, and a mathematics class with Colombian adolescents. The likelihood for the three episodes to move into landscapes of investigation is discussed. Aspects that could have led these classes away from the path of landscapes of investigation are highlighted: these include the demand of the school institutions for teachers to follow a specific teaching material and/or regulate class time, students' movements to reinforce the culture of exercises, and the failure of education managers to legitimise work aimed at the development of democracy. The

> analysis of the episodes showed the presence of aspects that could contribute to the promotion and maintenance of democracy: the dialogical posture of the teacher, and both teachers and students' opportunities to discuss social, cultural, political and ethnic issues in mathematics classes.

This chapter deals with relations between aspects of democracy and different mathematics classrooms with the potential to become landscapes of investigation. The educational contexts are independent: a class on functions in Brazilian higher education, high-school students visiting a quilombola community,[1] and a mathematics class with Colombian adolescents. The idea for this paper emerged from the group of authors at the Second Colloquium in Research in Critical Mathematics Education, held in 2018 in Brazil. On this occasion, Ole Skovsmose gave a lecture on the erosion of democracy. According to Skovsmose (2019, p. 4), democracy concerns "procedures for voting; a fair distribution of welfare; equal opportunities and obligations; and rights to express oneself. Each of these ideals can be eroded". In Brazil, for example, the teachers' rights to express their philosophical and political perspectives have been eroded by the government's attempt to implement a project they call *"escola sem partido"* (non-party school) in which knowledge is supposed to be taught in a neutral way. This scenario has been worsening since the coup that removed President Dilma Rousseff from office in 2016. Since then, Brazil has experienced a climate of loss of educational rights, intolerance to differences, and protests by an extreme right-wing minority calling for the establishment of a dictatorship.

Inspired by Skovsmose's lecture and by the political context experienced in Brazil during this period, the group discussed the links between democracy and mathematics classes. During this discussion, we observed that when working in mathematics classrooms with the potential to become landscapes of investigation, the concepts of mathemacy, dialogue, representativeness, equity, inclusion,

---

1   Quilombola refers to the communities formed by the descendants of African men and women brought to Brazil as slaves. By the Lei Áurea, law adopted on 13 May 1888, they came to own the land they lived in. This type of social organisation originally symbolised the black resistance struggle against slavery and the Portuguese colonial domain in Brazil.

collaboration and criticism become essential for the maintenance and promotion of democracy. Then, we wrote relevant episodes based on our own projects and classes.

The first episode, titled "But, Are We Not Going to Go Through the Conceptual Explanation First?" details a dialogue about functions and graphics between a teacher and students in a mathematics class in Brazilian university education. In the second episode, entitled "A Visit to the Quilombola Community Colônia do Paiol," we describe a pedagogical investigation developed by high school students (from 15 to 17 years old), participants of a research group in their own school. Finally, in the third episode, called "Democratic Abstention, Not Always a Favourable Choice", we report a situation in a mathematics class in Bogotá (Colombia) which dealt with an important issue in students' lives.

After presenting the episodes, we will explain aspects of democracy in mathematics classes, especially in environments that show the potential to be landscapes of investigation.

Episode 1: But, Are We Not Going to Go Through the Conceptual Explanation First?

It was Thursday night. The students of the Business Administration course of a higher education institution were waiting for the arrival of the mathematics teacher (one of the authors of this text). Around 7pm, she walked into the room, greeted everyone, and said:

Hi, folks! Are you all excited to review functions and graphics? Do you remember the classes that dealt with these subjects when you were still in high school?

I remember those x and y lines, one student said.

Oh, this has to do with those tables we used to build graphs, another complemented.

Some students said that the subject was too complicated and that they could never understand it properly. Others said they had already done many exercises on the subject. One student said she liked the topic. However, another colleague pointed out that he had always hated it, and stressed that functions and graphics did not "get into his head" and that he had not even had mathematics classes. The teacher listened attentively to each position taken.

I don't really understand why we have to learn all this, and this way! We don't even use it in everyday life... said one of the girls.

And it's worth adding that when something like this comes along, we can do everything using technology. Today it's all computerised! And in our area (Business Administration) we will hardly ever use mathematics the way we learn here, another colleague added.

The arguments they used seemed to want to justify or legitimise something already standardised by the traditional teaching of mathematics, the teacher thought. They addressed a few more discussions on those issues, and the teacher suggested that the students indicated some contexts that involved the use of functions.

But, are not we going to go through the conceptual explanation first? The one on page 43 of the handout? asked a student.

It's true, teacher! I myself have even completed some exercises on domain and image in advance... I remembered the definition... another student said.

The teacher talked about the use of the handout to support classes and praised the students' interest; however, she reinforced that she had another goal, to make them think about the use of functions in non-academic situations. Thus, the teacher and the students mentioned some examples: the relationship between the sales of a product and the salary of a person who takes commission on it; the values related to the variation of household bills, such as electricity; and the time spent to get to college that depends on different factors.

The teacher let students freely speak up. As they spoke to her, she wrote keywords on the blackboard to register each example. Then, she made a graphic representation of traffic peak times in the city. Here, the teacher used information the students gave her based on their daily experiences. They mentioned traffic jams in São Paulo, radars and Uber dynamic hours.

Then, together, they created a problem-situation about the calculation of the value obtained by a taxi driver after a given ride. The students suggested five reais for the fixed and initial value of the ride, and three reais for each kilometre travelled. Then, the teacher asked the students to say some values for the distance of the ride in order to calculate its price. After that process, the teacher represented the fictitious problem-situation created by the function $P(x) = 5 + 3x$, where x indicates the total of kilometres travelled and $P(x)$ the final price of the ride.

But are we going to solve the exercise first? asked a student.

No, we're not gonna solve it! We will explore how all these points you mentioned are beyond the Cartesian plane, the sketch of the graphic... look at the keywords in the middle of the board! Watch your lines, watch your routines. You have dealt here with various issues: means of transport and their impact on the quality of life. We could use mathematical concepts to further understand the relationships between the number of unemployed and emerging professions in society, such as the performance of Ubers, for example. We could comprehend how stressful situations can jeopardise the quality of life. Understand how public policies relate to the installation of radar and traffic violence rates. Would it be possible to understand what we study with questions like these? Think about it. Think about the ways we can interpret these tables, these graphs, these functions that, in the view of many, we use "so little".

The teacher allowed the students a few more moments to reflect, and they gave several contributions. Then, she opened the much-desired page 43, mentioned by some students at the beginning of the class. There were opinions about the issues raised. Meanwhile, from the middle of the room, a voice was heard:

It was kind of nice arguing about these things, too bad, unfortunately, we're a little behind on the content... it's already 7:47 pm.

Episode 2: A Visit to the Quilombola Community Colônia do Paiol

This episode is part of a broader pedagogical programme, entitled Social Research Group (GPS, in the Portuguese acronym). This proposal has been under development for more than ten years in two public schools and, although it is not formally integrated into the curriculum, it is traditionally part of mathematics activities in these school institutions. Student participation is voluntary. The objective of the project is to bring social themes that are important to democracy into the context of mathematics classrooms.

The described activity occurred in 2018 in one of the project schools, involving twelve students in their second and third years of high school.

Here we refer to them as *student-researchers*. This stage of the project included research visits and studies in a quilombola community known as Colônia do Paiol, located in the municipality of Bias Fortes (state of Minas Gerais, Brazil). In this proposal, the professor-coordinator (one of the authors of this text) and the twelve student-researchers discussed the ethnomathematical knowledge and artefacts that were part of the daily lives of the residents of that community, their narratives and life perspectives.

The preparation for the visit included several study meetings. These studies aimed to familiarise student-researchers with the characteristics that "anthropological" profile work should have: politeness in the treatment given to respondents; care not to invade the residents' privacy, and, especially, respect for their manifestations, beliefs, values and political positions.

The first trip to Colônia do Paiol was on a Saturday, August 2018, and the participating group was composed of the student-researchers, the teachers, and the school management team. The bus that took the group arrived in the quilombola community around 9am. The group went straight on to the community school, where the president of the residents' association and the school teacher were waiting for them. In the same day, the school was hosting a June party, and this was interesting, as it served to accommodate the student-researchers. They were introduced to the students and teachers at the community school and then dispersed throughout the school premises. Organised into four small groups (G1, G2, G3 and G4), the student-researchers talked and conducted semi-structured interviews with all those present in the school: teachers, students, coordinators and lunch ladies.

G1 recorded the first interview with Maria,[2] the school's pedagogical coordinator, who reported certain challenges experienced at school and in the community. Aspects more directly related to ethnomathematical knowledge and artefacts emerged during G2's conversation with Joana, the president of the residents' association. She coordinated and mediated the meetings between the student-researchers and the group of residents. She also provided information on the types of work—performed mainly

---

[2] The names we present in this episode are fictitious to preserve the anonymity of the people involved.

6. Aspects of Democracy in Mathematics Classes   101

by men—in the community. She told us that most of the men worked in the Serraria[3] or on the "courts".[4]

> The court is measured with a stick, now, I don't know how many metres long that stick is. Paulo will tell you [...] it's a bamboo stick. Then I do not know if they measure or if it is just by looking, that they are already used to, they go there in the woods... they calculate that court there... I remembered! There is even an expression, they say it like this, "the 12-stick court."

To learn more about this work, G2 interviewed Paulo, the foreman,[5] following Joana's suggestion. He said that the larger the area, the greater the number of workers he needed to recruit. The smaller the area, the smaller this number would be. The size of the region to be weeded is determined by using bamboo sticks as the unit of measurement. In this activity presented by Paulo, a stick was five metres long and, in general, the courts are squares measuring ten sticks on the side. The foreman explained to the students how the work was done. He talked as he drew a picture on the floor, illustrating how they measured the terrain.

> A court must be ten sticks upwards... but it's all square. Let's suppose [Drawing a picture on the floor] that's ten here, ten here, ten here, and ten here. So each court of five sticks, that is, five metres long, it counts like this five times ten, 50. It means that the ten would be 500m each side, so let's suppose if you join the ten sticks here, it would be two thousand metres, right!!!

The student-researchers took notes on the way of measuring presented by Paulo and continued conducting interviews throughout the morning of that Saturday. In the afternoon, there was a meeting where these and other reports were presented and shared with all participants in the project. Around 6pm that same Saturday, the groups ended their work in the community and returned home.

---

3   Name of a company and the region where it is located.
4   Working in the "quadro" or "quadras" (TN—translated here as "courts") refers to weeding, or reaping, in the woods of farms in the community region and in other states, such as Rio de Janeiro and São Paulo.
5   To weed, or reap, the courts, landowners hire a foreman (a person who organises and recruits a group of workers for the weeding).

Episode 3: Democratic Abstention, Not Always a Favourable Choice

This episode deals with a mathematics class for adolescents between the ages of 14 and 16 in a school on the southwestern outskirts of the city of Bogotá, Colombia, in 2016. This episode is reported by the class teacher, who is also one of the authors of this article.

At first, the teacher focussed on working through algebraic factorisation exercises. However, some students began to comment that factorisation was too easy a subject and that the teacher should discuss something more important for their social lives in that class.

The teacher, knowing how important the students' proposal was, accepted the challenge. She took advantage of the political situation in Colombia at the time. It was a country with more than fifty years of armed conflicts, still experiencing a war between the government and the Revolutionary Armed Forces of Colombia (FARC, in Spanish acronym). She brought to the classroom a conversation about the referendum that would take place the following week to get the public's opinion on a peace agreement between the government and the FARC.

Students were interested and commented that each should vote yes or no. The population's position should be summed up as "yes", in favour of a dialogue between the government and the FARC for a peace agreement, or "no", to keep the war going.

The students were aware that the political leaders had requested the agreement at the time: these were President Juan Manuel Santos and FARC leader Rodrigo Londoño, known as Timochenko. They also knew that the pact had been drawn up in Havana, the capital of Cuba, and that it was a complex document of almost 300 pages. The points of this agreement divided public opinion.

The Colombian teacher, born during the war, questioned her students about their opinions of that democratic event, the referendum. She said she had seen many Colombians die. According to the newspapers, the number was already over 260,000 dead, thousands missing, and more than eight million victims in general. She asked them what they were thinking about that first peace deal.

The students commented that their parents showed no interest in voting, but that they were against the war. They also pointed out that they all knew and no longer endured so many conflicts, quarrels, and violence.

The teacher and students talked about the referendum and discussed how important it was that everyone expressed their choices at the polls, taking a stand for or against the agreement. The teacher wanted students to think about the relevance of voting, since this was a constitutional right acquired in the country.

The students got increasingly interested in the subject and assured the teacher that they would talk to their parents and ask them to go to the polls to vote in that referendum. After this dialogue, the teacher returned to the explanation of the exercises involving algebraic factorisation, since the school would demand the accomplishment of the content planned for that class. However, she agreed with the students that the discussion would be resumed in another opportunity.

The referendum took place the week after that class. For everyone who participated in the discussion, the result of the election was a great surprise, as the "no" to the peace agreement between the government and the FARC did win. Of the valid votes, 50.2% represented "no" to the peace agreement, and 47.7% expressed a "yes". Abstention was 60%, which corroborated the students' opinion on the population's non-participation to referendums.

The result saddened the students because even though they knew that most of the people they knew were against the war, they realised that it made no difference, since they had not gone to the polls. The students concluded that the great absence of these people for voting contributed to the continuity of the armed conflict that had been devastating the country for more than five decades.

4. The Interconnection between Democracy and Landscapes of Investigation

The classroom is an environment in which different students meet to learn. During this process, they interact with each other, with the teacher, and with other participants in the school context. Such interactions may contain democratic or authoritarian aspects. The classroom can thus be understood as an environment in which students learn to obey rules and adapt to society, or as a micro-society in which they learn daily to relate to each other, to respect and learn from differences, and to make decisions.

Skovsmose and Valero (2012) refer to ideas of collectivity and transformation, and highlight how intrinsically related these aspects are to democracy. Thus, it is possible to establish relationships in the classroom that can support the maintenance and promotion of democracy.

Paulo Freire (1959, 2014) stresses the importance of the political dimension of education. Literacy has the fundamental role of providing conditions for students to understand themselves as human beings who can participate in the society in which they live and modify it. The educational process should thus have as one of its essential objectives the aim of creating conditions for the student to use knowledge to recognise and reflect on situations of oppression that are present in society. Such reflections can cause students to engage in actions that fight for change.

Skovsmose (1992) critically discusses the political dimension of mathematical knowledge and its contributions to the development of a democratic posture in the classroom. According to the author, mathematics learning has a technical dimension—which refers, for example, to the learning of procedures and algorithms—but also a political dimension. To encompass these two dimensions of mathematics learning, Skovsmose used the concept of mathemacy, which is closely related to Paulo Freire's previously mentioned idea of literacy. Similarly, besides calculating skills, mathemacy enables reflection on situations in which mathematics is involved, explicitly or implicitly, allowing us to question these situations and engage in actions aimed at change. Such situations are called "mathematics in action" by Skovsmose (2014). In developing mathemacy, we make a critical reading of situations, evaluating the positions of power, the risks involved, and the possible changes.

Providing an environment to develop mathemacy in mathematics classes requires political and democratic posture from the teacher. It requires teachers to leave their comfort zone to enter a risk zone where they cannot control every future event. Penteado (2001) characterises a risk zone as a territory of uncertainty and unpredictability. This uncertainty, however, can provide an environment of new learning for teacher and students.

In the comfort zone, the teacher has in mind the steps to be taken by the class and expects the students to follow them. In this context, characterised by Skovsmose (2001a) as a paradigm of exercise, the teacher mostly dominates the speech. In this environment, there is no room for students to take responsibility for the learning process or to converse about social and political approaches.

Thus, to develop mathemacy, the classroom cannot be configured only as a space where we solve exercises, in which bureaucratic absolutism that "states what is right and wrong in absolute terms" usually stands (Alrø and Skovsmose, 2004, p. 11), without explaining the criteria that guide such decisions. The teacher's role in this environment is to identify and correct students' mistakes, based on the absolute truths of mathematics. For example, the kind of error made—whether an algorithm was incorrectly used, or a statement was misinterpreted—is not discussed. There is no possibility of questioning the data of an exercise. The dominant form of communication in this context, especially between the teacher and students, is a quiz. Usually, the questions are asked by the teacher, and it is he/she who approves or disapproves of the students' answers. These answers are often not discussed. In this classroom, students can learn to use mathematical techniques and algorithms, which is a type of learning that agrees with the paradigm of exercise discussed above.

When considering the ability to criticise mathematical information and reflect on its meanings, another type of communication and interaction between teacher and students is necessary. They must discuss what is proposed, understand different perspectives, actively listen to what the other says, and reformulate ideas. These are the aspects that characterise dialogue in mathematics education (Alrø and Skovsmose, 2004; Milani, 2018; Faustino, 2018). A teaching and learning environment to develop mathemacy is dialogic. Such an environment can be provided in landscapes of investigation.

Skovsmose (2001a) presents the landscapes of investigation as a favourable learning environment for the development of mathemacy.

> A landscape of investigation invites students to formulate questions and to look for explanations. The invitation is symbolised by the teacher's "What if ...?" The students' acceptance of the invitation is symbolised by their "Yes, what if ...?" In this way the students become involved in a process of exploration. The teacher's "Why is it that ...?" provides a challenge, and the students' "Yes, why is it that ...?" illustrates that they are facing the challenge and that they are searching for explanations (Skovsmose, 2001a, p. 125).

For students to engage in this learning environment, they must feel invited, and this depends on the quality of the invitation. "A landscape only becomes accessible if the students actually accept the invitation. The possibilities of entering a landscape of investigation are dependent on relational qualities" (Alrø and Skovsmose, 2004, p. 50). We cannot ignore that landscapes of investigation may be attractive for one group of students, but uninteresting for another.

When teachers and students are engaged in landscapes of investigation, the form of communication that emerges is dialogue, which is characterised by dialogic acts that constitute the Inquiry Cooperation Model (Alrø and Skovsmose, 2004): getting in contact, locating, identifying, advocating, thinking aloud, reformulating, challenging and evaluating. According to the authors, these elements, empirically observed in mathematics classes, may not be the only possible dialogical acts. They assist both in maintaining and developing dialogue.

By putting these dialogic acts into action, teacher and students might establish a democratic relationship, since everyone has the right to speech and active listening, and the understanding of what each participant says is fundamental (Milani, 2018). By active listening, we mean what Alrø and Skovsmose (2004, p. 62) propose: "asking questions and giving non-verbal support while finding out what the other is getting at". Working in landscapes of investigation requires that students agree on and decide which paths to follow in the activity. Listening to each other, understanding what each other says, reformulating their ideas, considering their perspective, and evaluating the work done are democratic aspects of dialogical interaction.

The theoretical elements discussed above will contribute to our emphasis and discussion about aspects related to democracy in the three episodes of the mathematics classes.

5. Aspects of Democracy in Mathematics Classes

The first aspect to be discussed is how dialogue appears in the three episodes. In Episode 1, the teacher tried to demonstrate a dialogical posture in relation to the students by calling them to participate in the class with their experiences and opinions, by valuing their comments, listening carefully to what they said, and by using their contributions to build the class. This attitude of the teacher characterises the active listening necessary for the dialogue to develop. This search for students' participation in the construction of the class characterises a democratic attitude. Not only did the teacher have a voice, but this was shared with the students. Both parties' perspectives were highlighted and clarified. The teacher sought to discuss aspects that went "beyond the Cartesian plane", while some students sought actions related to the paradigm of exercise. The clarification of perspectives contributed to the democratic classroom environment that the teacher was trying to establish.

In Episode 2, student-researchers visited a culturally different community than the one which they were used to. The teacher who coordinated the visit had already discussed with them the attitude of respect they should assume regarding the knowledge, beliefs, values and political positions of the community members they would visit. We did not feel that there was an imposition of knowledge, either on the part of the visiting group or that of the residents of the community, which would characterise what Freire (1967) calls cultural invasion. The students did not invade the quilombola community culturally, but rather opened themselves to communication, guided by a posture of respect for the other to learn new ways of how mathematical knowledge can come into action, such as in foreman Paulo's talk about the use of sticks to measure the courts. Guided by a dialogical posture of respect, tolerance, active listening and understanding the other as a source of knowledge, the students experienced a learning environment that recognises differences and values them, that realises that there are different ways of being in the world, and mathematises it. According to Frankenstein (2012, p. 53),

> when people respect each other's intellectual activity, many taken-for-granted ways of our own thinking and knowing are challenged, thereby expanding students' (and teachers') interests and imaginations. I believe that we have an ethical responsibility to 'force' our students to grapple with a progressive analysis of what is going on in our world.

Thus, the democratic aspect related to dialogue is also constituted in the recognition and appreciation of differences, enabling the identification of the other as a source of learning.

Concerning Episode 3, the dialogue that took place in the classroom made it possible to create a horizontal relationship between the involved. The teacher could talk about her anguish and what she went through because of the conflicts in her country. Everyone could listen carefully to what was said, in a relationship of respect or exchange of experiences, trying to understand the different perspectives of the result of that referendum. We consider a democratic posture on the teacher's part as realising that the participation of students in the construction of the class is fundamental, since the mathematics class space consists of all of them, with their interests and individual learning processes.

Beyond the dialogic posture described above, we can stress another element from the three episodes. According to Skovsmose and Valero (2012, p. 9),

> a mathematics education that is committed to democracy cannot be reduced simply to the intrinsic qualities of mathematics, or to the conceptual constructions of the discipline. Instead, there are many social, political, economic, and cultural factors that must be taken into account as constantly directing and redirecting their development.

In Episode 1, the teacher's dialogue with students on aspects that go "beyond the Cartesian plane" had the pedagogical objective of discussing how mathematics can influence daily issues that are part of people's lives. In Episode 2, the visit, also considered to be a mathematics class, we dealt with diverse social and cultural themes. The project, which the episode is part of, contributes to education on racial relations. Students and people from the quilombola community exchanged knowledge from their own cultures and tried to understand different points of view. It also enables students to develop different interdisciplinary perspectives on the same theme. Skovsmose (2020) has drawn attention to the need for ethical and political aspects to be addressed in mathematics classes as well. The author warns us of the danger of the banality of mathematical expertise in a democracy. According to Skovsmose (2020, p. 7), "a banality of mathematical expertise refers to the high priority of being able to do mathematics compared to the capacity of being able to reflect on the ethical impacts of what one is doing".

Episode 3 reinforces how important socio-political discussions are for the mathematics classroom. Engaged students can promote reflections on critical social issues, such as elections. Classroom situations like this might enable the development of a democratic competence (Skovsmose, 2001b); that is, a minimum set of knowledge that citizens need to have in a representative democracy to ensure its functioning. The interaction between teacher and students illustrates the importance of discussing the relevance of citizens' votes as a mechanism of choice. It is a decision-making process that can lead citizens to participate in political processes and practices that contribute to a critical attitude towards their actions in society. The action of questioning the citizens' low participation rate in elections began in that specific mathematics class.

These mathematics classes indicated that both purely mathematical topics and "issues related to everybody's lives" or "beyond the Cartesian plane" could contribute to different qualities of mathematics learning, emphasising the democratic aspects of a class with regard to the type of discussion generated.

The episodes described do not configure landscapes of investigation, as Skovsmose (2001a) defines them. The discussions between teachers and students in Episodes 1 and 3, and between student-researchers and members of the quilombola community in Episode 2, suggest a starting point for entering landscapes of investigation.

The class reported in Episode 1 ended up focussing on the study of intrinsic aspects of mathematics, such as domain and image of a function and its graphical representation. Even with the call of some students for this study, the teacher could, in fact, have entered a landscape of investigation, deepening one of the aspects raised when reflecting with the students, such as means of transport. What are the advantages and disadvantages of using Uber? Who wins and who loses with this use? What is the difference in price between taking a taxi ride and Uber? How does mathematics help us understand this difference? What is the impact of sharing an Uber ride with someone else on the final value of the ride? These are questions that could have been explored in that mathematics class, approaching a landscape of investigation.

The visit to the quilombola community Colônia do Paiol reported in Episode 2, also considered to be a mathematics class, was part of a research project that involved students in the discussion about racial and mathematical issues. After the visit, one possibility for developing investigative work at school would be to discuss and try to understand the way to calculate the courts, explained by the foreman, Paulo. What mathematics do they use to measure the courts? Shall we try to reproduce Paulo's explanation in drawing? What other ways of measuring do you know that could be applied in the case of the courts? Why are there different ways to measure? Further contact with the organisers of the visit to the quilombola community could clarify the relationship between the wages the workers received and the areas they reaped, their working hours, and the comparison between the wages received by the foreman and each worker. Thus, beyond the study of different ways of measuring, other mathematical concepts could contribute to the understanding of the relationships involved in the context at stake: function, a simple rule of three, proportionality, area of plane figures and perimeter. Mathematics would figure as a tool to read and write the world (Gutstein, 2006).

In Episode 3, the mathematics teacher begins the class by dealing with algebraic factorisation. Subsequently, she meets the request of some students to discuss more important issues for their lives. Finally, she resumes the algebraic study they had approached at the beginning of the class. The continuation of the discussion requested by the students could have followed a different path. Some questions could have been raised: what would the trajectory of the teacher and students have been like if the conflict between the government and the FARC, which began well before they were born, had not occurred? Why did people decide not to go to the polls to vote on the referendum if they longed so much for peace? Or did they long for war? But how do we know what they want if they do not go to the polls to vote? Should everyone prioritise living in peace? Why is voting mandatory in some countries? And if it is a democracy, why is voting compulsory? How has non-participation been in the Colombian elections? What do statistical and graphical data tell us about it? What is the impact of the abstention rates on the outcome of the elections? These questions represent the possibility of developing literacy (Freire, 2014) and mathemacy (Skovsmose, 2001a) in mathematics classes, approaching a landscape of investigation.

We imagine these future possibilities for the reported three episodes based on the dialogical posture of the teachers involved in each of these mathematics classes. Fostering a discussion about how mathematical functions relate to students' everyday issues, conducting a visit to learn how other social groups measure objects, and developing a discussion about the importance of voting in elections all represent essential starting points for teachers and students to enter landscapes of investigation. This can be interpreted as an aspect of democracy in the pedagogical stance of the three mathematics teachers.

Why did the classes not follow a path close to the one imagined and described earlier? What prevented the teacher and the students from entering a landscape of investigation? We will discuss this aspect, considering each context in which the episodes occurred.

The educational institution of Episode 1 was a private university that had its own handouts and established the order of content that should be handled in class. Such determination directly influences the distribution of content throughout the classes. The questions raised by the teacher in class were not included in the didactical material. Thus, class time does not belong to the teacher and students, but to an external factor: the teaching material of the university. Another aspect of this episode that contributed to investigative work not advancing was the comments of some students, who were familiar with the traditional practice of mathematics education. We can observe this when they reveal their concerns about doing the exercises and studying the mathematical content on page 43 of the textbook. But are we not going to go through the conceptual explanation first? But will we solve the exercise first? These remarks show how relevant this practice seems to be to students: using the correct procedures, determining the solution of the exercise, and obtaining the teacher's confirmation of the answer. This sequence of actions is the one we find in the paradigm of exercise, characterised by bureaucratic absolutism.

In this episode, we see two movements towards different places in a mathematics class: traditional mathematics education, based on a sequence of actions characteristic of the paradigm of exercise, and guided by some students and an environment that approaches the beginning of a landscape of investigation, highlighting the teacher' actions in a risk zone (Penteado, 2001) when trying to create a democratic environment.

Episode 2 occurred in the second half of 2018, at the same time that the state electoral process in Minas Gerais occurred. The result of that process was the transfer of political power from a left-wing party to a neoliberal one. One of the consequences for the work that was being carried out in the school was the interruption of the project in 2019. The work followed informally, but it was no longer legitimised by the management.

In Episode 3, the students challenged the teacher to discuss an issue that had not been planned: the referendum. She accepted the challenge and, at the end of the class, resumed the study of the mathematical content initiated. As we presented in the analysis of Episode 1, the teacher knew that she needed to continue with the study of the content, because she would be questioned by the school's pedagogical coordinator. It is clear to us that the teachers had a greater or lesser degree of freedom, depending on the institutional context in which they are inserted, to make their choices (change the progress of the class, for example). But the fact is that breaking with bureaucratic absolutism and questioning what is already regulated does not seem to us an immediate process.

In this analysis, we highlight aspects of democracy related to mathematics classes presented in three episodes. From what has been described in those classes, we imagine possibilities for teachers and students to enter landscapes of investigation with a dialogical posture as a reference. Finally, we highlight some possible reasons that led the classes described not to follow the path into such landscapes.

6. Final Considerations

The idea to write this text arose from the discussion group formed by the authors at the Second Colloquium in Research in Critical Mathematics Education, based on Ole Skovsmose's lecture on erosions in democracy. Our interest was strongly influenced by the struggle against the authoritarian relations that plague Brazil. These situations of oppression are increasingly present in our society, and they are also entering the classrooms.

In this article, we presented three episodes of mathematics classes, lived by the authors themselves. Our goal was to illustrate democratic aspects in such scenarios.

The analysis of the episodes showed the presence of aspects that can contribute to the promotion and maintenance of democracy in the classroom environment. One of the aspects concerns the dialogic posture of the teacher by inviting students to participate in the class, actively listening to what is evidenced by students, valuing and legitimising what is said, and providing moments of recognition and appreciation of differences. Furthermore, one can observe an attempt by both parties to understand the perspectives presented, and a shared responsibility between students and teacher for the construction of the class. In this way, the dialogue could become a democratic form of interaction between the teacher and the students.

Another aspect of democracy raised was teachers' and students' opportunities to discuss social, cultural, political and ethnic issues. Both such questions and issues intrinsic to mathematics provide possibilities for working with mathematical concepts. This was reinforced when we presented the likelihood for the three episodes to move into landscapes of investigation.

We also highlighted aspects that may have led these classes away from the path towards landscapes of investigation. The demand of the school institutions for teachers to follow specific teaching material and/or regulate class time was highlighted in two of the episodes. Our analysis also underscored certain students' movements to reinforce the culture of exercises and the pattern of traditional mathematics classes. Also, education managers did not legitimise a work aimed at the development of democracy.

We consider that there are different kinds of resistance to offering students the opportunity to explore social problems in mathematics classes. However, this is a challenge which should be faced by schools and teachers. The responsibility for change is shared by students and teachers from all school disciplines. The mathematics teacher can contribute to creating a place for relevant discussions for citizenship and democracy, besides establishing an exercise resolution environment.

References

Alrø, H., & Skovsmose, O. (2004). *Dialogue and learning in mathematics education: Intention, reflection, critique*. Kluwer Academic Publishers.

Frankenstein, M. (2012). Beyond math content and process: Proposals for underlying aspects of social justice education. In Wager, A. & Stinton, D. W. (Eds), *Teaching mathematics for social justice: Conversations with educators* (pp. 49–62). USA: The National Council of Teachers of Mathematics.

Faustino, A. C. (2018). *"Como você chegou a esse resultado?": O diálogo nas aulas de matemática dos anos iniciais do Ensino Fundamental*. Doctoral dissertation. Universidade Estadual Paulista (Unesp).

Freire, P. (1959). *Educação e atualidade brasileira*. Tese de concurso para a cadeira de História e Filosofia da Educação na Escola de Belas Artes de Pernambuco. Recife.

Freire, P. (1967). *Educação como prática da liberdade*. Paz e Terra.

Freire, P. (2014). *Pedagogia do oprimido*. 58. ed. Paz e Terra.

Gutstein, E. (2006). *Reading and writing the world with mathematics: Toward a pedagogy for a social justice*. Routledge.

Milani, R. (2018). "I am sorry. I did not understand you?": The learning of dialogue by prospective teachers. In J. Moschkovich, D. Wagner, A. Bose, J. Rodrigues & M. Schütte (Eds), *Language and communication in mathematics education* (pp. 203–216). Springer.

Penteado, M. G. (2001). Computer-based learning environments: Risks and uncertainties for teachers. *Ways Of Knowing Journal, 1*(2), 23–35.

Skovsmose, O. (1992). Democratic competence and reflective knowing in mathematics. *For the Learning of Mathematics, 12*(2), 2–11.

Skovsmose, O. (2001a). Landscapes of investigation. *ZDM—Mathematics Education, 33*(4), 123–132.

Skovsmose, O. (2001b). *Educação matemática crítica: A questão da democracia*. 3rd edition. Papirus.

Skovsmose, O. (2009). Preocupações de uma educação matemática crítica. In M. H. Fávero & C. Cunha (Eds), *Psicologia do conhecimento: O diálogo entre as ciências e a cidadania*. (pp. 101–114). UNESCO, Universidade de Brasília, Liber Livros Editora.

Skovsmose, O. (2014). *Critique as uncertainty*. Information Age Publishing.

Skovsmose, O. (2019). Inclusive landscapes of investigation. *Proceedings of the ninth international mathematics education and society conference*. MES10.

Skovsmose, O. (2020). Banality of mathematical expertise. *ZDM Mathematics Education, 52*(6), 1187–1197. https://doi.org/10.1007/s11858-020-01168-4

Skovsmose, O. & Valero, P. (2012). Rompimiento de la neutralidad política: El compromiso crítico de la educación matemática con la democracia. In P. Valero & O. Skovsmose (Eds), *Educación matemática crítica: Una visión sociopolítica del aprendizaje y enseñanza de las matemáticas* (pp. 25–61). Uniandes.

7. Collaborative Learning within Critical Mathematics Education

Bülent Avcı

> This chapter investigates ways in which collaborative learning in critical mathematics education can promote critical citizenship and democracy. Drawing on critical participatory action research in a US high-school classroom, the article argues that the critical mathematics education approach to collaborative learning is a coherent alternative to neoliberal, market-driven approaches. The results suggest that collaborative learning within critical mathematics education should aim to transfer classrooms to communities of learners in light of dialogic pedagogy and inquiry-based education.

Collaborative learning is a pedagogy that sets regulative norms, through which students study and learn together. In a classroom context, it can take different forms and be oriented toward different educational goals. Historically, collaborative learning was a critique of traditional approaches, but it later diffused into mainstream pedagogy and curriculum, and lost its original orientation. Today, it is often practised in mathematics classrooms, where it is regarded as a necessary component of effective pedagogical strategies.

The current rationale for collaborative learning is that research has found that it increases students' learning. For example, drawing on classroom-based research, Pietsch (2009) concludes that collaborative learning is an essential part of effective mathematics teaching. Goos (2004) argues that collaborative learning makes learning more meaningful—less a matter of rote memorisation—and, moreover, produces cultural forms of learning. Carpenter and Lehrer (1999) suggest that collaborative learning creates a learning environment where collective-public reflection can emerge. Franke, Kazemi, and Battey (2007) assert that collaborative learning creates a discursive ambiance that helps students learn mathematics better. Johnson and Johnson (2016) consider collaborative learning in relation to citizenship and democracy, as it promotes interdependence among peers and helps them to develop the values, attitudes and behaviour of engaged citizens. In theory, collaborative learning seems to have the potential to help students develop values and skills in order to become critical and engaged citizens. Here, it is necessary to frame collaborative learning in a way that cultivates a "thick" version of citizenship and democracy, as described by Wright (2015): that is, as justice-based, participatory democracy. A citizen in this view is someone who is committed to equality and freedom; someone who possesses a sense of collectivity, solidarity and empathy; someone who is actively involved in the decision-making process and can question authorities and hold them accountable; someone who can come together with other citizens to take civil initiatives based on common interests and concerns. Such a citizen is different from the "neoliberal" citizen who is law-abiding and individually responsible, but who practises a form of citizenship ultimately rooted in consumerism.

Collaborative learning toward critical citizenship within critical mathematics education (CME) has not been studied as a prime unit of analysis to date. CME can be seen as a subset of critical pedagogy. It is a relatively new but growing domain of research concerned with social, cultural, and political implications of mathematics education (Avcı, 2017, 2018; Gutstein, 2006; Skovsmose, 1994, 2011; Skovsmose and Greer, 2012). Theoretical studies in CME claim that mathematics education can be oriented toward critical citizenship and democracy (Skovsmose, 2011). Thus, construction of CME as a research field requires defining basic educational concepts, such as collaborative learning, so that they are aligned with the central concerns of CME. As a mathematics teacher and researcher, the question of how collaborative learning in CME can be distinguished from its mainstream, neoliberal[1] versions preoccupied

---

1 Mainstream literature driven by neoliberal ideology often uses the concept of collaborative learning in business contexts: students (workers) should be able to

me for a long time. More specifically, I wanted to understand what the word *critical* in CME signifies.

Unfortunately, there are only a few classroom-based studies that reject the neoliberal definition of collaborative learning (Avcı, 2017). As Skovsmose (2014) suggests, the word problems in the current study aimed to move away from the exercise paradigm and in the direction of *landscapes of investigation* in order to counter neoliberal hegemony in education. To address this gap, this chapter draws on classroom-based data to investigate the ways in which collaborative learning within CME can counter neoliberal pedagogy[2] and promote critical citizenship and participatory democracy.

1. Methodology

Life in the classroom is a complex, interactively dynamic, context-dependent phenomenon, and so understanding it requires a qualitative approach that "capture[s] the complexity of being human in an interacting social world of teaching" (Pine, 2009, p. 17). Thus, research methodology should resonate with the objectives of study in terms of epistemology and ontology in order for us to understand the dynamics of collaborative learning. Therefore, I adopted an action research methodology because it is well suited to the classroom. Resonating with the natural flow of classroom teaching, action research allows the cycle of plan-act-observe-reflect (Carr and Kemmis, 1986). The methodology enables students to democratically participate in classroom activities and the process of knowledge construction. The methodology can be considered an adaptation of critical participatory action research (CPAR), as framed by Kemmis et al. (2014), in which students are seen as active agents of change as opposed to passive objects of the process: research is done not *on* students but *with* them. Therefore, the methodology is conceptually consistent with the concerns of CME (Avcı, 2020).

---

      work as a team to complete duties assigned by a teacher (employer). CME should reclaim collaborative learning as an agency of change.

2   It is the dominant approach to education in the USA, which considers teaching as a *science* rather than an *art* and claims education is neutral—it can thus be standardised and delivered through 'best practice' and measured by standardised testing. It illustrates students as customers and teachers as salesmen/women.

Action research allows researchers to intervene at different stages. As Herr and Anderson (2005) state: "Unlike traditional social science research that frowns on intervening in any way in the research setting, action research demands some form of intervention" (p. 5). According to Carr and Kemmis (1986), action research is a form of self-reflective inquiry undertaken by participants (teachers, students, or principals, for example) in social (including education) situations in order to improve the rationality and justice of: (a) their own social or educational practice, (b) their understanding of these practices, and (c) the situations (and institutions) in which their practices are carried out (p. 162).

This definition of action research resonates with the objectives of my research. Habermas (1972) argues that knowledge and human interests are strictly correlated, and that the desire for knowledge entails three categories: technical, practical and emancipatory. Drawing on Habermas's theory of communicative action, Kemmis et al. (2014) relate each interest to a unique epistemological stance and consequently to its own research methodology. Accordingly, types of action research can be categorised as technical, practical and emancipatory. As Kemmis (2009) points out, these three types "involve very different kinds of constellations of sayings, doings and relating" (p. 469). In CPAR, which is emancipatory, we "find and enact ways of doing things that are less irrational [...] less unproductive [...] and less unjust or exclusionary" (Kemmis et al., 2014, p. 68).

I designed the current project as a case study that draws on CPAR's plan-act-observe-reflect methodology. The study was conducted in a high-school mathematics classroom where I teach full-time. It involved a year-long, everyday mathematics (pre-calculus) class with thirty-two students, aged fourteen to seventeen. Data were collected from student journals and whole-class discussions, as well as from my own field notes and reflective journal. As Winter (1989) recommends, I analysed the data from a dialectical perspective, which features three basic premises: (a) although it is a unified whole, a phenomenon is structured in relation with other phenomena; (b) phenomena exist in a context that contains opposite (uncoupling) forces; (c) phenomena are in constant change: change is the result of tension between the unity of the phenomenon and the uncoupling forces. In order to align the research with these dialectical premises, I used the "versus" coding method developed by Saldaña (2013). Saldaña's system adequately captures the tensions between collaborative learning in CME and market-driven education.[3]

2. Preparation Stage of Lessons and Projects

As I was developing this project, I realised that there were many factors to consider. For years, the mathematics curriculum in USA public high schools has been standardised; indeed, it is admired for being "teacher-proof". Micromanagement policies are in place that tell teachers what, when and how to teach. On the other hand, because CME is a "subversive" form of teaching, teachers who wish to practise CME must find ways in which it can be situated within the standardised curriculum. Accordingly, I developed projects (lessons) and integrated them into standardised units. In this way—and despite the control that filtered down from district and local administrators—I was able to practise CME during regular school hours.

Collaborative learning should be a process through which students can develop democratic values and attitudes. As Johnson and Johnson (2013) argue, it would be self-defeating to impose democratic values from the top down. Instead, collaborative learning should allow students to develop agency by organising their own groups and making other changes to democratise life in the classroom.

Prior to developing the action research projects, I had some questions and concerns:

- Problems and questions should not be too easy or too difficult. In both cases, students may lose concentration or be discouraged. Students should need mathematical content knowledge to participate in group discussions.

- Projects should help students improve their content knowledge; this is necessary to counter a widespread bias against critical pedagogy where critical pedagogy (and CME) are thought to be inconsistent with rigorous learning.

- Peer interaction in small groups should be non-dominating and dialogic. However, this runs counter to the USA school

---

[3] Market-driven education and neoliberal education-pedagogy are used interchangeably.

system, where learning is structured as competitive. How to challenge this fundamental belief? Should we have whole-class discussions to discuss the pros and cons of collaborative vs. competitive learning?

- Should students set their own groups or should I be involved in the process? In the USA, ethnic pride is often confused with multicultural education. What if students set groups based on ethnicity or race? This would defeat our purpose, as collaborative learning should help students interact with their peers based on universal values, regardless of race or culture.

As I developed projects, I needed to be aware not to fall into default, neoliberal ways of thinking. For example, word-problem themes should not be based on competition or consumerism, but rather should resonate with ethics of emancipation, collectivity and solidarity; they should also relate to students' own life-worlds. This can motivate students so that they do not need extrinsic motivational support.

Each project needs to be linked to the state standards. As micromanagers, school administrators often visit classrooms to ensure teachers follow the standardised curriculum. I needed to make the grade.

As we became immersed in the first project, my students and I gradually began to find answers to these questions. In cycles of plan-act-observe-reflect, some questions evolved into others, and new concerns emerged along the way.

3. Critical Participatory Action Research in the Mathematics Classroom

The study included five projects, one per month (Table 1). Two were mechanical exercises; the others were open-ended word problems. Each project required two lessons; each lesson is a ninety-minute-long block period, with the first lesson done on the first day and the second lesson on the next. We began each with a whole-class discussion concerning objectives. One of our goals was to achieve egalitarian peer interaction and collaborative learning. I then introduced the project; this was followed by group work and group presentations. Then we had another class discussion to reflect on our practices. At the end of the project (Avcı, 2019), students made entries in their journals. I took field notes and wrote in my reflective journal.

Table 1. Content and themes of end-of-unit projects (EUPs).

EUP	Mathematics Content	Theme
1	Linear equations and functions	Standardised assessment
2	Multipart functions: fishing town	Critical mathematical literacy
3	History of mathematics	Universal values of humanity
4	Systems of equations and inequality	Community service and justice
5	Exponential functions	Student loan debt crisis

Findings led to the conclusion that collaborative learning in CME may be distinguished from neoliberal (mainstream) pedagogy. As Table 2 indicates, CME and neoliberal pedagogy differ across five key domains.

Table 2. Key differences between CME and neoliberal pedagogy.

Domain	Critical Mathematics Education (CME)	Neoliberal Pedagogy
1. Setting up the groups	Students set and self-organise their own group in an egalitarian manner; sense of community and horizontal relationships	The teacher assigns each student to a specific task or responsibility; negligible sense of community
2. Motivation for learning: extrinsic vs. intrinsic	Intrinsic; subject matter relates to students' life-worlds (e.g., pollution)	Extrinsic (test scores, grades); subject matter typically from consumer or business world (e.g., best-buy or profit maximisation option)

Domain	Critical Mathematics Education (CME)	Neoliberal Pedagogy
3. Power dynamics in peer interaction (ZPD)	Interactions of equals	Replication of existing dynamic: the teacher is the authority; peer tutoring may lead to emergence of banking model
4. Learning materials	Open-ended word problems afford links to larger world and multi-layered responses	Skill-drill exercises limit opportunities for empathy and collaboration
5. Classroom life: democracy vs. compliance	Justice-based participatory democracy; "thick" version of democracy and citizenship	Collective compliance; individually responsible students carry out specific tasks; "thin" version of democracy and citizenship

4. Results and Discussion

4.1 Setting up the Groups

Setting up groups for collaborative learning may seem to be an insignificant detail; however, my research revealed that group-setting is a crucial step, as it directly relates to the goals of collaborative pedagogy. Without having any discussion of group-setting, I let students set their own groups for the first two projects. My intention was to observe students' natural inclinations towards group work. For these projects, students seemed to come together mostly based on cultural and ethnic background. Eduardo voiced this sentiment in his journal:

> It was easy for me to say, "Hey, James, let's do this [project] together". They are my friends anyway, but I am not sure what the other dude would say about it.

This quotation is from the first project,[4] which indicates that the class was not yet a community: students chose their friends, not ready yet to work together regardless of background. As the teacher, I noticed that there is a dialectic relation between the classroom being a community and successful application of collaborative learning. Collaborative learning is something that has to be learned through systematic work.

Organising groups to be multicultural and cross-gender was challenging at first. However, as we went further, students seemed to be interacting, studying and learning better in group work. During the third and fourth projects, we discussed setting up groups, and agreed that we should be able to work with all classmates, regardless of ethnic, racial or any other identity—this is our classroom, our community. In the fifth project, students set their own groups again. But this time, it was noticeable that the groups were, by any measure, much more mixed.

In a larger context, students' tendencies to feel comfortable with others from their own background was not a coincidence. It is an observable fact that the power elite in the USA uses racial, ethnic, religious and cultural differences to keep people from coming together to organise themselves around their common interests and problems to take collective initiatives in the public arena. Therefore, a practice of CME aimed at promoting critical citizenship should encourage students to develop civic skills to work collaboratively with their peers from diverse backgrounds.

After reflecting on previous experience with group work, some students proposed that assigning each member a specific task would improve the quality of work. This was another important question: should the teacher assign each member of a group a specific task? We had several whole-class discussions on this issue, and agreed that assigning each student a task defeats the purpose of collaborative learning, which is to help students develop the skills, attitudes and values of engaged and critical citizens. Whereas collaborative learning should give students autonomy and help them grow as independent learners, assigning them specific tasks would be controlling and micromanaging. Similarly, when people (citizens, workers, activists, etc.) organise grassroots movements, it is usually a matter of civil initiative. No one is officially (top-down)

---

4   The theme of this project was the standardisation of curriculum and assessment; students develop multiple-choice assessment in small groups (Avcı, 2019).

assigned a specific task—citizens define problems, develop plans and mobilise themselves. They may perhaps decide on some division of labour, but that is a joint decision, not an imposed one.

It is interesting to reflect that small-group collaboration in which each member is assigned a specific task is, however, functional from a business perspective: employers seek workers who can work with others to carry out specific tasks assigned by the manager. But such pedagogy would be far from helping students to develop a sense of solidarity, collectivity and empathy. Instead, top-down tasks would appear to promote individually responsible, consumer-based citizenship. Indeed, if a person completes their part of the work without having any say in the overall problem, then this process may produce alienated individuals who do as they are told to earn a wage (in the business context) or credit (in the educational context). This situation is not collaboration but compliance.

Therefore, collaborative learning in CME should empower students to set groups with peers not necessarily from their ethnic group, to determine division of labour, and to complete group work democratically. Working towards egalitarian collaboration in the classroom is not to deny that individual students have different talents and skills. Through the five projects, there was no evidence that a student considered their peers to be obstacles to success. We established a synergistic relationship between individual students and groups such that individual growth led to growth of the whole classroom community and vice versa.

4.2 Motivation for Learning: Extrinsic vs. Intrinsic

Students were asked to reflect on their group-work experiences in the past. Lauren articulated her frustration: "This is what happens. One or two do all the work and everyone else in the group gets credit from that work without doing their fair share." This is a legitimate concern when learning is structured within a reward/punishment paradigm. Students' journals also indicated that when assignments do not relate to their own life-world, some complete the project just for the credit, while others do not even bother with the credit.

From the perspective of market-driven education, Farrell and Lawrence (2016) consider the kind of collaboration we practised as counter-productive to the purpose of career readiness. Their position is a reminder that pedagogical practices involving cooperation and collaboration are never neutral, but have contentious ideological and political implications. Market-driven education—which aims to promote consumer-based, individually responsible citizenship—promotes collaborative learning in school, because it envisions students as future workers who are going to need skills to work with others to solve problems assigned by managers.

At the start of each project, we also debated whether it should be graded. It was decided that learning is its own reward: we did not need academic credit for every activity undertaken in class. Therefore, the projects were not graded in the traditional sense. However, I took special care to ensure that the projects were related to students' life-worlds.[5] Thus the motivation was intrinsic, not extrinsic. Many students, like Jennifer, expressed their opinions in their journals:

> I don't care if it was graded or not, the fishing-town project was a really enjoyable one. It made me feel like we were working on something real... It was not like doing a couple of math exercises.

As Jennifer indicates, it is important that themes of projects in CME should be derived from students' life-worlds—something they find interesting. In this way, collaborative learning in small groups can be engaging and meaningful for students.

4.3 Power Dynamics in Peer Interaction (ZPD)

Vygotsky's (1978) notion of the zone of proximal development (ZPD) was one of my inspirations towards collaborative learning in the first place. According to ZPD theory, students' intellectual performance differs drastically when performing alone compared to when performing with the assistance of the teacher or a more competent classmate. However, surprisingly, my experience led to the conclusion that ZPD in the context of peer tutoring may produce power relations among students, generating what Freire (2000) referred to as the "banking" concept of education. During the first project, as I observed group work in which one student was teaching the other, I realised it

---

5   Whether students find pure mathematics interesting and relate it to their life-worlds or not should be another research study. For teaching and learning pure (mechanical) mathematics, dialogic pedagogy and inquiry-based approaches distinguish CME from mainstream education (Avcı, 2019).

was an example of typical transmission-style education; there was an observable vertical relation between the two students. This may not be an issue in mainstream (neoliberal) education. However, if the objective is to foster a culture of collectivity and solidarity, peer tutoring must not be allowed to become another version of transmission-style education. If students cannot interact with each other as equals, then the classroom is not an egalitarian community. ZPD might help students learn mathematics better and improve their functional literacy; however, from the perspective of CME, it has the potential to be a dehumanising and oppressive learning experience.

Students also identified self-assigned leaders as obstacles to egalitarian work. The power of self-appointed leaders can come from a variety of sources. I noticed that if students can bring their status with them into the classroom, peer interactions tend to reproduce power relations and hierarchy. Then one student dominates the other members of the group. During the first two projects, I observed a particular student, a popular football player, who was dominating the group: the other members seemed to be passive followers, and their voices went unheard. In a different group, a student whose parents were lawyers (high status in a low-income school) interacted with her group in a way that clearly indicated that the group was supposed to follow her directions.

We hold whole-class discussions to identify obstacles to and opportunities of democratic peer interactions. Before and after each project, we discussed ways to achieve egalitarian collaboration. For example, students indicated that self-assigned leaders in small groups work as an obstacle to egalitarian peer-interactions. Students' reflections also indicated that they actually enjoyed interacting with each other as equal members. As we completed more projects, students began approaching one another with empathy. Denny's journal entry for the fifth project captures this sentiment:

> I always thought I knew the answer...I figured it out before anyone else, but when we talked with each other, it was much more than what I thought...There were many different ways to calculate how much the oil company should pay fishing families.[6]...I realised how important it is to listen to your friends and see their thoughts and feelings.

---

6   "Fishing Town" was one of our projects: a big oil company polluted the ocean, and citizens sued the corporation for compensation.

Thus, as time went on, egalitarian collaboration gradually emerged. Students made a conscious effort to establish empathetic and non-dominating interactions with their peers. It would be very difficult, if not impossible, to create a learning environment where students could learn with and from each other if power relations or hierarchies of any kind played defining roles in interactions. Pedagogy oriented towards collaborative learning to promote participatory and justice-based critical citizenship requires the classroom to be an egalitarian community where students' interactions are guided by empathy and solidarity.

4.4 Learning Materials

Five cycles of planning, teaching, observing and reflecting on collaborative learning led to the conclusion that learning materials and projects must be conducive to collaboration. This point is directly aligned with the notion of *landscape of investigation* developed by Skovsmose (2014). Three of five projects involved open-ended word problems that provided communicative space for students to link their learning to the larger world outside the classroom. In these problems, there was no single correct answer; these were inquiry-based, multi-layered questions that allowed students to negotiate different approaches. The following quote is from one of the small-group projects where students worked on the real-world problem that deals with optimisation:

> *Nadia*: If we maximise something, don't we need a quadratic function?
> *Nicole*: I think so...but some linear inequalities can be optimised. You know, just like we did in exercises last week from the textbook...
> *Nadia*: We need to maximise a function, but what function?
> *Nicole*: Yeah...that is what we need to figure out...We need to write that equation down first...Here we need to add family and individual unit prices [writes inequalities down].
> *Tom*: Why did we set inequalities in standard form and put into slope-intercept later?
> *Nicole*: It is easier to graph it in slope-intercept form.
> *Tom*: Then we should have set it in slope form in first place.
> *Nicole*: I don't know how to set it. It becomes easy to set it in standard form when I read the problem... [They write all inequalities and the objective function down].

Tom: I wonder if Edward[7] gets money for this work.
Nadia: Maybe some pocket money...Why not?
Nicole: He is doing a community service here; he has a part-time job...you know, in the other project he got a part-time job.

Once students agreed that the system of inequality corresponds to Edward's story and the objective function, they moved on to calculations and graphing:

Nadia: Looks like our solution area has three edge points...x-intercept is 20 and y is 15.
Nicole: I got the same, but let's plug them in and see how they work....I also got the intercept of two lines.
Tom: Yes, x and y intercepts are correct.
Nadia: Let's evaluate objective function...Tom, can you do it by calculator?
Tom: Yes, P(x,y) is the objective function right? [Pointing to their objective function]
Nicole: Oh yeah, that's the equation [she points to the equation]. This is going to calculate the total money they could get.

This quotation reveals that each member of the group contributed to the process of collective thinking. Even though their skills and knowledge varied, they learned from and with each other.

The other two projects involved practice sessions on mechanical exercises. Mathematics contains properties, axioms and theories, etc. Comparing skill-drill exercises to open-ended word problems, the exercises seemed not to provide a lively communicative space. Students working on exercises in groups appeared to be less motivated and more likely to digress. This outcome is supported by Skovsmose (2014) in his article "Landscape of Investigation", who writes that CME should be critical of the official curriculum—he conceptualises it as an exercise paradigm and creates a landscape of investigation for students to be engaged in meaningful learning processes. The landscape of investigation in my research formed a communicative space that provided a solid ground for dialogical pedagogy and collaborative learning.

In this context, I openly communicated the following point with the students: in today's market-driven education system, where the success of schools, teachers and students is measured by standardised test

---

7   Edward is the character in the word problem who volunteers in a community service.

results, it is important for students to have functional literacy and be able to pass the tests. That is to say, students need a skill-drill type of mathematical knowledge to be successful in a traditional sense. However, this does not mean that teaching mechanical exercises in mathematics is a neutral process where the pedagogic approach makes no difference. For teaching the mechanical aspects of mathematics, CME should make a distinction between authoritarian and dialogic pedagogy: mathematics can be taught and learned through dialogic pedagogy. Anything that can be taught through the authoritarian approach can also be taught through dialogic pedagogy. Of course, CME strongly prefers dialogic pedagogy as it cultivates democratic values and attitudes.

4.5 Classroom Life: Democracy vs. Compliance

Throughout five projects, as we worked to achieve egalitarian collaboration, our classroom became a community of learners. In this process, I, as the teacher, became a facilitator. Students' interactions with each other became less dominating and more empathetic; classroom life became more democratic. Such effects demonstrated that our discussions of justice-based participatory democracy and critical citizenship were not empty rhetoric. The more that collaborative learning democratised the class, the more participatory became our discussions on ways to transform the class into a community. For example, by the time we completed the third project, the following list of norms had been agreed on:

- In group work, we will approach each other empathetically, not judgementally. We have a common goal to achieve.

- We are continuously working to establish non-dominating interactions. Regardless of where we come from, we are all equal in this class; we have an equal right to say things and do things in group work.

- We actively participate in discussions in our groups and in the class. We focus on individuals' arguments, not on their power or personality.

- If for some reason a member is shy and unwilling to participate, group friends welcome and encourage them.

- We can disagree on things; that is normal. However, it is important that we democratically solve these disagreements without excluding anyone. A group member may have a different way of solving problems or answering questions. We solve issues through open dialogue and egalitarian peer interaction.

- Nothing should be imposed on group members: groups decide how to plan, carry out and present the project.

Agreeing to these regulative norms meant that students made a public commitment to them. Students developed a noticeable sense of belonging as individuals contributed to group work and found that their ideas and work were valued. Unlike the situation that exists in competitive learning, our collaborative activities did not create winners and losers. Quite the contrary: collaborative learning processes dialectically connected individual students' success to the success of the whole class. As such, the collaborative learning process was an important element in transforming the class into an egalitarian community of learners.

With respect to the connection between democracy and mathematics education, Dewey (1916) claimed that scientific education, including mathematics, leads naturally to a democratic society, because science and mathematics reject external authority of any kind. My research, however, led to a different conclusion: how mathematics is taught is equally, if not more, important than the content itself. If mathematics is taught in an authoritarian way, it will promote authoritarian, not democratic, values. I agree with Skovsmose and Alrø (2004), who suggested that in this respect CME needs to transcend the views of Dewey.

In conclusion, the study showed that high-school mathematics students can interact with one another in non-dominating, dialogic, and empathetic ways. We found that collaborative learning in the context of CME could be structured as a process through which students experience learning with and from each other and become agencies of change to democratise life in the classroom. Collaborative learning is a sustainable practice when it gradually transforms the class into an egalitarian community of learners. Such learning promotes the values and skills that students need to become active and critical citizens instead of producing or reproducing existing power relations. Thus, collaborative learning in CME is sharply distinguished from collaborative learning in mainstream educational practices; CME collaborative learning runs counter to neoliberal pedagogy.

I hope it is apparent from the foregoing that this was participatory action research in which students were, to a large degree, active participants and co-researchers. Loren's final journal entry sums up her thoughts and feelings:

> To be honest, I am not a math person... Math was never my strength. I always felt anxious as I stepped into my math classes...but this year was quite different. I mean like *really* different. We were all talking together in class to improve our group work, make our class an equal place...I was included in all projects we did. Everybody asked questions of each other and we talked about a lot of social issues and learned a lot...In the fishing-town project...I actually learned how to find intercept points of lines and circles with and without a calculator...I will miss this class.

References

Alrø, H. & Skovsmose, O. (2004). *Dialogue and learning in mathematics education: Intention, reflection, critique.* Kluwer Academic Publishers

Avcı, B. (2017). *Critical mathematics education in neoliberal era.* Doctoral dissertation. Charles Sturt University.

Avcı, B. (2018). *Critical mathematics education: Can democratic mathematics education survive under neoliberal regime?* Brill-Sense.

Avcı, B. (2020). Research methodology in critical mathematics education. *International Journal of Research & Method in Education, 44*(2), 135–150. https://doi.org/10.1080/1743727X.2020.1728527

Carpenter, T. P. & Lehrer, R. (1999). Teaching and learning mathematics with understanding. In E. Fennema & T. A. Romberg (Eds), *Mathematics classrooms that promote understanding* (pp. 19–32). Erlbaum.

Carr, W. & Kemmis, S. (1986). *Becoming critical: Education, knowledge, and action research.* Falmer Press.

Dewey, J. (1916). *Democracy and education: An introduction to the philosophy of education.* Free Press.

Farrell, J. P. & Lawrence, G. (2016). *Rotten to the (common) core: Public schooling, standardized tests, and the surveillance state.* Process.

Franke, M. L., Kazemi, E. & Battey, D. (2007). Understanding teaching and classroom practice in mathematics. In F. K. Lester (Ed.), *Second handbook of research on mathematics teaching and learning* (pp. 225–256). Information Age Publishers.

Freire, P. (2000). *Pedagogy of the oppressed.* Continuum.

Goos, M. (2004). Learning mathematics in a classroom community of inquiry. *Journal for Research in Mathematics Education, 35,* 258–291.

Gutstein, E. (2006). *Reading and writing the world with mathematics: Toward a pedagogy for social justice.* Routledge.

Habermas, J. (1972). *Knowledge and human interests.* Beacon.

Herr, K. & Anderson, G. L. (2005). *The action research dissertation: A guide for students and faculty.* Sage.

Johnson, D. W. & Johnson, F. P. (2013). *Joining together: Group theory and research.* Allyn & Bacon.

Johnson, D. W. & Johnson, R. (2016). Cooperative learning and teaching citizenship in democracies. *International Journal of Educational Research, 76,* 162–177.

Kemmis, S. (2009). Action research as a practice-based practice. *Educational Action Research, 17*(3), 463–474.

Kemmis, S., McTaggart, R. & Nixon, R. (2014). *The action research planner.* Springer.

Pietsch, J. (2009). *Teaching and learning mathematics together: Bringing collaboration to the centre of the mathematics classroom.* Cambridge Scholars Publishing.

Pine, G. J. (2009). *Teacher action research: Building knowledge democracies.* Sage.

Saldaña, J. (2013). *The coding manual for qualitative researchers.* Sage.

Skovsmose, O. (1994). *Towards a philosophy of critical mathematics education.* Kluwer Academic Publishers.

Skovsmose, O. (2011). *An invitation to critical mathematics education.* Sense Publishers.

Skovsmose, O. (2014). Landscapes of investigation. In O. Skovsmose (Ed.), *Critique as uncertainty* (pp. 3–20). Information Age Publishing.

Skovsmose, O. & Greer, B. (2012). *Opening the cage: Critique and politics of mathematics education.* Sense Publishers.

Vygotsky, L. S. (1978). *Mind in society: The development of higher psychological processes.* Harvard University Press.

Winter, R. (1989). *Learning from experience: Principles and practice in action-research.* Falmer.

Wright, D. E. (2015). *Active learning: Social justice education and participatory action research.* Routledge.

8. Global Citizenship

Manuella Carrijo

> The word citizenship, a contested concept, emerges in different discourses that can hide contrasting interests. Differences in its meanings are enhanced with the changes in society through history. This chapter presents a discussion about mathematics education for citizenship and its outcomes. School presents itself as a stage of dispute and a space for expressing different ideologies that can configure ways of thinking and structuring an education on the topic of citizenship. Mathematics education is shown to be supportive of different citizenship discourses. In a globalised world with deep inequalities, a type of mathematics education for citizenship that matches with inclusion and diversity is required; one which considers global issues. The proposed landscape of investigation *Global Visibility Matters* is a possible support aid for mathematics classes, based on enabling students to develop global citizenship. By reading and interpreting a social situation as being open to change, it becomes possible for the students to be recognised as—and to act as—global citizens.

Mathematics education for citizenship is a matter that causes a lot of ambiguity. The term *citizenship* has been used with very different connotations and interpretations. "It has become a dangerously consensual word, an empty envelope into which the dreams of a society of equals, a society of rights and duties can fit, as well as a society divided by antagonistic interests" (Gadotti, 2010, p. 67, my translation). Accordingly, it is a term that can be applied to very different contexts.

In Carrijo (2014), I analyse theses and dissertations in mathematics education and draw attention to the importance of moving beyond the

© 2022 Manuella Carrijo, CC BY-NC 4.0   https://doi.org/10.11647/OBP.0316.08

generic way of using citizenship in educational discussions. I find it important to present an epistemological position without falling into an empty discourse and using citizenship as a buzzword.

In our globalised world with intense inequalities, precarious public services, social and political instability, contexts of violence, and social and economic vulnerability, any desire of building a more humanised world based upon the notion of citizenship meets huge challenges.

Extensive social changes also cause changes in school. Mathematics inserted in the power scheme of a globalised context also influences educational practices and interpretations of citizenship education. In this context, it is important to explore mathematics education for citizenship as an educational concept that not only advocates for democracy or human rights, but also includes a global perspective. But what are the notions underneath a mathematics education for citizenship? What are the possibilities that working with landscapes of investigation might provide for further development of an education for citizenship?

This chapter aims to reflect on citizenship education and how it might relate to mathematics education. First, a brief historical review of the concept of citizenship, including a political and social panorama, will be presented. Subsequently, a discussion will be raised about the context of mathematical education, and how its organisation and teaching objectives relate to the promotion of citizenship. Next, global issues will be addressed, making it necessary to broaden this discussion. Finally, an inspiration for landscapes of investigation is proposed as a possibility for teaching mathematics for global citizenship.

1. Citizenship: A Contested Concept

The concept of citizenship is very volatile and dynamic. It follows a historical process, changing as society changes and also as ideologies change. To talk about citizenship is to pass through a discursive minefield, in which generic uses of this word can be related to completely different interpretations. Therefore, it is necessary to indicate what kind of citizenship one refers when using this concept.

The word *citizenship* has a contested meaning. According to Figueiras et al. (2016), a contested concept can be understood as a concept without any well-defined meaning. It can be given different interpretations and they can operate in very different discourses. Furthermore, a contested concept represents controversies that can be of profound political and cultural natures. This should not be taken to imply, however, that attempts should be made to prevent the use of contested concepts. It is precisely their contested nature that makes it possible to facilitate discussions, where agreements and disagreements move a dialogue forward (p. 16).

The understanding of citizenship should not be seen as an apolitical and uncritical matter. Human interests and purposes are based on ideals with respect to belonging and non-belonging, and of favouring and disfavouring. Those ideals are sometimes veiled and reflect the need for a deeper understanding of what citizenship might include.

The term *citizenship* originated when cities that believed in the importance of discussing and determining the rights and duties of individuals emerged. Its emergence correlates with the demand in ancient societies for popular participation in the destinies of the collective (for example, by voting). Later, ideas about the sovereignty of the people, individual freedom and legal rights had intense influences on many other societies. Concerns about human privileges, guided by arguments based on divine law and the right to ownership of private land, were also related to the understanding of what it means to be a citizen worldwide.

Through the demise of feudalism and the rise of capitalism, the Industrial and French Revolutions of the eighteenth century were made possible. New forms of state organisation were established in relation to individuals, societies and state apparatus. The idea of legally equal citizens was introduced, although inequalities in economic capacities were maintained. The appearance of citizens with properties and citizens without properties foreshadowed bourgeois class society and replaced ideas based on predestination and religious reasons for acceptance of inequalities (Guarinello, 2003). This context gave rise to a new vision related to the development of science and mathematics.

Recently, our understanding of citizenship has been strongly influenced by economic globalisation. In the face of neoliberalism, the free mobility of capital around the world and the reduction of state intervention in the economy (minimum state) have provided room for social wealth privatisation and profitable exploitation. The responsibilities of the state have been transferred to the individual. Consequently, the precariousness of public services and social inequalities has intensified (Heywood, 2010). Thus, there emerges a citizen governed by economic patterns and guided by efficiency, productivity and competitiveness in a reality formed by merit, with money as a unit of measure. Globalisation and neoliberalism are plainly not beneficial to everyone on the planet. This is evidenced by the widespread manifestation of exclusion and social inequality.

In Brazil, for example, the understanding of citizenship is likewise based on a set of cultural, political, socio-historical and economic transformations. It has intensified during times of European exploitation and has been shaped by the slavery and domination of Indigenous and African people. It has passed through dictatorship and authoritarianism, resulting in political instability. Such developments are directly linked to limited and fragile access to citizenship.

According to Gomes (2003), since the time of colonisation, social interaction between people of different ethnicities has been based on historical experiences of cultural conflicts, domination, protests and confrontations, arising from diverse identities. This scenario shows the process which influenced the establishment of a huge part of Brazil's marginalised population.

With the growing complexity of societies and the many socio-political changes, the concept of citizenship has often been used. This might give the impression that the concept has some universal meaning across these contexts. However, this is not the case; rather, it appears that citizenship has become a buzzword.

Benevides (1994) and Beitz (2009) warn that some interpretations may be considered to be *false citizenships*. Examples could be: *Citizenship Tutored*, when social relationships are conducted only by the market, with little state intervention, and social policies are characterised by demobilisation; *Passive Citizenship*, when citizens do not have or do not exercise democratic political power over the state and all rights are at the consent of an authoritarian government; *Regulated Citizenship*, where the exercising of citizenship depends on a specific profession or productive activity, regulated by law; and *Assisted Citizenship*, where social relations are regulated by the market, with the mediation of a controlling state, with welfare policy aimed at expanding social rights (but not political rights).

In this context, citizenship may have a market-oriented version that assumes that everybody is equal, apart from their particular capabilities. From this perspective, one can consider inequalities with respect to wealth, social position and political power as originating from the talents and skills of the individual. Hence, opportunistic discourses claim that education is necessary in order to gain skills and techniques that are relevant first of all for building up capabilities and, as a consequence, for the labour market. This way, school has a questionable function of serving the interests of capital by operating as a company with education as its commodity. In particular, mathematics education becomes related to production processes and to some ideological actions, making it necessary to update schools in line with new economic imperatives.

Benevides (1994) and Guarinello (2003) argue that it is possible to consider the notion of citizenship from four perspectives that differ and are complementary: the first is *Citizenship and State*, where the term "citizenship" is used as a synonym of nationality. In Brazil, for example, people with Brazilian nationality are considered Brazilian citizens, and this idea is one of the criteria used by legislation to differentiate rights and duties. The other perspectives are *Citizenship and Human Rights*, which relates to certain rights—such as civil, political and social rights—being minimum conditions for existence, which must be guaranteed to all people in society; *Citizenship and Democracy*, which concerns citizen exercising different political attributions and is related to the awareness of rights and duties; and *Citizenship and Society*, which is about a collective vision that establishes common values and goals shared by the whole society, and is implemented in favour of the community or the nation, although contrary to more immediate individual interests.

I consider the concept of citizenship as reaching beyond the synonym of nationality. It is critical to go beyond nationalists' conceptions of citizenship which are employed to justify borders and exclusion. I am in favour of citizens of the world, without distinction of race, sex, language or religion. Thus, I consider the discussions of human rights, democracy and society to be important issues when talking about education for citizenship.

Facing this complex framework, many demands arise with respect to social injustices. It becomes important to foster a sort of citizenship that considers inclusion and challenges inequity. Social movements, for instance, fight for reorganisation in order to achieve the minimum dignified living conditions. Demands about poverty, migrant people, the rights of women, black and indigenous people, and the struggles of the LGBTQIA+ community are some of the topics that need to be addressed.

Education based on "false citizenships" can help to retain the privileges of the bourgeoisie through unequal access to quality education, since access to education is an important aspect of fostering democratic and active political participation. Santiago and Akkari (2020) argue that:

> It appears evident that conservative and elitist governments are reluctant to unlock the full potential of schools to form active and responsible citizens capable of building a more just egalitarian society in which different cultures and a plurality of epistemological knowledge coexist (p. 23).

Education and citizenship are intertwined. Faced with a reality of inequality at the global level, questions about the conditions of the least favoured and vulnerable population must be on the education agenda. This is also a challenge for mathematics education. In the next section, I will address how mathematics education plays a significant role in preparing students for citizenship and how important it is to consider the different understandings of this contested concept.

2. Some Citizens Are More Citizens Than Others

In the book *Animal Farm*, Orwell (1945) writes political satire about a group of farm animals who rebel against their human farmer. They intend to build a society where the animals can be equal and free. But even though at first the animals apparently create an equal society, the pigs in charge apply their power to oppress the other animals through manipulated information and fear. In the end, the principle *All animals are equal* changes to *All animals are equal, but some animals are more equal than others*. Similar kinds of modifications of principles might be applied to human situations, and as in the way described by Orwell, reproduce the oppressive behaviour of humankind.

It is possible to connect Orwell's fable with education. Since citizenship is a precondition for possessing certain rights—and those considered less worthy citizens are not granted these rights—education can be used to maintain and justify hierarchies in citizenship. Education operates in a space of contradictions, which means that it assumes an institutionalised social order. However, it also provides ideas for changing that order.

Thinking about the connection between mathematics and society, Pais (2012) states that some discourses place mathematics in the realm of powerful knowledge, providing competencies necessary for becoming a full-fledged citizen and worker. In a context marked by the sciences and technologies, in which much information is connected to mathematical language, mathematics becomes established with even more power. As a consequence, people without this knowledge might be prevented from participation in social affairs. They may be considered less capable, less productive, and therefore lesser citizens. Therefore, it becomes essential to reflect on the responsibility and influence of mathematical knowledge for building citizenship.

Mathematics can influence discourses about the role of people in society and validating citizenship. This is due to the fact that mathematics education has great power in enabling or preventing people from accessing certain positions in society, as well as the roles they might come to play in the workplace. The gate-keeping function of mathematics reaches much further than access to further education; it concerns access to social possessions and welfare in general. The failure to understand mathematics can lead to a range of limitations that might cause people to renounce their rights.

In addition to mathematics knowledge, we should draw attention to the valorisation that mathematics gives to people is also something to draw attention to. It is not a "problem of skills or knowledge, but a matter of what is valorised in our social symbolic order" (Pais, 2012, pp. 65–66). Why does mathematics have an important and prideful place in school? Why is it used as evidence in international educational assessments like PISA (Programme for International Student Assessment)? It is certain that this power of mathematics has to do with economic and technological knowledge, since these topics are highly valued in a globalised world.

Skovsmose (2020) points out that by forming patterns of legitimation and justification, mathematics is brought into action with an impact on decision-making processes in society. Thus, there is a trust in the presence of mathematics that is established as an integral part of authority in society. Mathematics can consequently format society, and this requires considering an ethical dimension in the face of a range of social implications. Altogether, these attributes might give power to a group of people that can master mathematics. It can make them greater citizens than others, since they are considered more capable of making decisions.

However, mathematics is not a prerequisite to becoming citizens. The point is not only the guarantee of "providing meaningful mathematical learning to pupils but guaranteeing that even if someone fails in school mathematics, they will be no less a citizen and no less able to participate in society than those who succeed" (Pais, 2012, p. 66). It does not mean, however, that it is not necessary to consider mathematics as a way of fostering citizenship. Mathematics plays an important role in dealing with social and technological issues, but this cannot be wrong-headed as a precondition for being a citizen or for creating a hierarchy between people.

Mathematics can be a resource that serves diverse kinds of interests and it is possible to have a mathematics education for the purpose of citizenship with different aims. According to Skovsmose (2005, 2008) and D'Ambrosio (2011), the situation is critical, mainly because of what has been produced with this knowledge. This includes both wonders and horrors.

One can consider mathematics to be a sort of knowledge that is difficult to come to master, that is designated for privileged people only. Thus, mathematics might be a tool to be manipulated in the face of the world of competition and the search for profits. Mathematics education can have the task of training citizens to accept the given order from the globalised information economy (Skovsmose, 2005). On the other hand, mathematics education can act as a promoter of citizens who are critical, aware and skilled in making decisions for the benefit of the global community. But how can mathematics education help in understanding people's places in a global world, and enable them to be autonomous-minded and willing to challenge social injustices?

3. Towards Global Issues

In 2015, the 2030 Agenda for Sustainable Development was featured, and all the member states of the United Nations approved a plan based on Sustainable Development Goals (SDGs) for accomplishing a better future for all.

> By 2030, ensure that all learners acquire the knowledge and skills needed to promote sustainable development, including, among others, through education for sustainable development and sustainable lifestyles, human rights, gender equality, promotion of a culture of peace and non-violence, global citizenship, and appreciation of cultural diversity and of culture's contribution to sustainable development (United Nations, 2015, p. 19).

According to Akkari and Maleq (2020), Global Citizenship Education (GCE) has taken a noticeable place in the 2030 agenda, which highlights the need to foster global citizenship in an increasingly interconnected world. In this sense, citizenship refers to belonging to a global community. For them, engaging in citizenship education includes considering demands for local and global challenges. The authors also emphasise that global citizenship is interpreted differently depending on the political, economic and cultural contexts. Thus, this concept based on human rights and peace is hardly likely to be addressed in a uniform way around the world.

Inspired by the GCE (Akkari and Maleq, 2020), I consider it crucial to promote mathematics education for global citizenship. Such an education has to enable students to understand global issues. It includes an awareness of other perspectives, a perception of oneself as part of the global community, and it helps develop a sense of social responsibility and solidarity towards less privileged groups of people.

Mathematics education for global citizenship must go beyond the walls of classroom and bring the students into the world around them by adding international issues to the programme. It is not about preparing the students to work productively in national or international economies, where mathematics skills are meant for economic productivity.

Mathematics education for global citizenship goes ahead to overcome the division between "us and them" and "here and there". Based on the principle of respect for diversity, it has to do with the commitment to human rights, environmental sustainability, peace and social justice, and exploring citizenship from excluded and marginalised people.

Such an education demands a critical and transformative approach. This means considering and respecting diversity in mathematics classrooms, ensuring inclusive and equitable quality education, and promoting learning opportunities for all. Such an education must also value diversities in mathematical knowledge, which means considering and learning together with others, and acquiring different kinds of mathematics knowledge. In the next section, I address a landscape of investigation, as inspired by mathematics education for global citizenship.

4. Landscapes of Investigation for Global Citizenship

The notion of landscapes of investigation is important for critical mathematics education. Landscapes of investigation are learning environments where students are invited to take part in an investigative process (Skovsmose, 2001). It is possible to create mathematics lessons where students have the chance to experiment, investigate and engage in dialogic processes. These landscapes create learning possibilities considering the interests of the students, and create room for autonomous and creative action.

Landscapes of investigation overcome the exercise paradigm, an approach of traditional mathematics teaching focussed on techniques necessary for solving exercises, with no space for questioning. According to this paradigm, mathematics lessons are guided by textbooks that, like exams, present exercises in a sequence of orders (find, calculate, demonstrate...). These exercises contain texts with little relevant information and with the sole purpose of framing the exercises. They have one and only one correct answer that, in most cases, can be found without any complementary critical or creative thinking.

For a mathematics education based on "false citizenships", it is enough to walk through learning environments based on the exercise paradigm. When based on the principles of efficiency, productivity and competitiveness, mathematics education promotes the development of skills for the globalised labour market, and not competences relevant for establishing global citizenship.

However, teaching mathematics through a landscape of investigation is in resonance with a global citizenship education. It is possible to include real contexts, explore, question and make proposals. This can help to create space for discussions about human rights and social reality, and students might come to engage in making social changes.

Britto (2013) planned mathematics lessons that relied on investigation into the visibility of black people in Brazil (see also Chapter 3, "Media and Racism" by Britto). Through a landscape of investigation, he proposed to the students in the final years of elementary school an investigation into the presence of black people in printed media. Students had to look for pictures of children in magazines and consider the context in which they were inserted. The students were asked to consider to what extent this type of media offered the same spaces to black and white people. As a result, the invisibility of black people was pointed out. Furthermore, in many pictures, they appeared in a negative situation.

This landscape of investigation provided students with an environment inviting them to carry out investigations and critical reflections by discussing racism, social inequality and stereotypes of black people. Indeed, this investigation provided room for students to discuss, by means of mathematics, essential matters with respect to citizenship.

Inspired by this landscape of investigation, it is possible to broaden the investigation toward GCE matters. Considering global issues, I briefly outline a landscape of investigation, which I refer to as *Global Visibility Matters*. In mathematics lessons, students might investigate, for instance, the extent to which the visibility of black people in international media is related to racism. One can consider the statistics not only in Brazil but in other countries as well. One can search through the images of black Brazilian people in international media, for example. How are pictures of black women represented in international media? Is it possible to identify sexualisation promoted by misinterpretations of some traditional parties like carnivals? How can the international portrayal of carnivals reinforce racism?

Dantas et al. (2021), for instance, analyse the covers of a famous fashion international magazine and present a chart that helps to understand the representativity of black women in this kind of international media. In *Global Visibility Matters*, one can also search for a list of covers of international magazines with black people on.

This landscape of investigation can help us to think about the underrepresentation or misrepresentation of black people. Students can come to consider how such patterns of representation might be projected into country relationships. In other words, this landscape of investigation can aid, through mathematical concepts, the interpretation of reality in a global context. The way the media reflects and enhances inequalities between countries—for instance, by giving less space to a group of people or showing them in a negative way—might legitimise and intensify social exclusion. Groups of people might be portrayed as being in miserable conditions, due to their own lack of abilities, and not due to some global oppressive forces.

By exploring a landscape of investigation, one might create space for action, which Biotto, Faustino and Moura (2017) call *landscapes for action*. Also, Gutstein (2006) points to the possibility of *reading and writing the world with mathematics*, where "reading" refers to interpreting the world, and "writing" to changing the world. Working with the landscape *Global Visibility Matters* might cause students to perceive themselves as agents of change and use mathematics to change their realities into a sense of belonging. An essential feature of a mathematics education for global citizenship is to broaden the ways for students to get involved in the problems of the school, the community, the country and the world. By reading and interpreting a social situation as being open to change, it becomes possible for the students to be recognised and to act as global citizens.

Landscapes of investigation provide students and teachers with the opportunity to leave their comfort zones and enter a zone of risk. This is an environment of uncertainty and unpredictability, with the emergence of unexpected situations (for a discussion of the notion of risk zone, see Penteado, 2001). In this sense, when students are exploring *Global Visibility Matters*, the teacher cannot predict what issues will arise and a risk zone is created. However, a risk zone is also a zone of possibilities; it might provide space for action.

5. Concluding Remarks

The term *citizenship*, a contested concept, appears in different discourses and can refer to contrasting interests. Such contrasts intensify with the changes in society over the course of history. In Brazil, these contrasts become denounced through manifestations of people in situations of vulnerability. People are deprived of their rights and taken to be the sole person responsible for their hardships.

School provides a space for manifestations of different ideologies that can configure ways of thinking and structuring education for citizenship. In this scenario of contradictions, the development of citizenship becomes a concern for critical mathematical education.

Living together in peace in a culturally diverse society is a basic for GCE. When coming together to do so, it is important to critically interrogate and challenge power relationships that reinforce domination and subordination between countries. Besides, it is imperative to enable students to assume active roles, both locally and globally, towards overcoming social injustices.

Mathematics education might help to shape people who can intervene in reality in different ways. It can support citizenship discourses directed towards productivity and competitiveness that can intensify violence, inequalities and social injustices. But mathematics education can also play an important role in promoting citizenship committed to global issues in an attitude of transformation towards social justice. The proposal of the landscape of investigation *Global Visibility Matters* indicates a possibility for mathematics education for global citizenship.

Mathematics education can be based on human rights, including fundamental anti-racism issues. It can engage in the current trends in international education policies to ensure a transition from "false citizenships" towards global citizenship. Surely, on its own, mathematics education cannot change the inequality in the world. But, as Freire (2000, p. 31) stated: "If education alone cannot transform society, neither will society change without it" (my translation).

Acknowledgements

This chapter is part of my PhD project, supported by the Coordenação de Aperfeiçoamento de Pessoal de Nível Superior (Capes) from Brazil and Ernst Mach Grant from Austria. It is also part of the Research Group Épura from Unesp, Rio Claro, São Paulo. I am grateful to Amanda Queiroz Moura, Célia Roncado, Daniela Alves, Denner Barros, Jammal Ince, Luana Oliveira and Peter Gates for helpful comments and suggestions.

References

Akkari A. & Maleq K. (2020). Rethinking global citizenship education: A critical perspective. In A. Akkari & K. Maleq (Ed.), *Global citizenship education* (pp 205–217). Springer. https://doi.org/10.1007/978-3-030-44617-8_15

Alrø, H. & Skovsmose, O. (2004). *Dialogue and learning in mathematics education: Intention, reflection, critique.* Kluwer Academic Publishers.

Dantas, I. J. de M., Soares, G., J., Batista, F. E. A., Cordeiro, R. B. & Silva, C. A. P. (2021). From subversion to diversity: The black representativeness on the covers of the American vogue magazine. *Educação Gráfica, 25*(1), 234–247.

D'Ambrósio, U. (2011). *Educação para uma sociedade em transição* (2nd edition). EDUFRN.

Benevides, M. V. de M. (1994). Cidadania e democracia. *Revista Lua Nova*, 33, 5–16.

Beitz, C. R. (2009). *The idea of human rights.* Oxford University Press.

Britto, R. R. (2013). Educação matemática e democracia: Mídia e racismo. *Annals VII, CIBEM*, 3355–3362. http://www.cibem7.semur.edu.uy/7/actas/pdfs/568.pdf

Biotto, D., F., Faustino, A. C. & Moura, A. Q. (2017). Cenários para investigação, imaginação e ação, *Revista Paranaense de Educação Matemática, 6*(12), 64–80.

Carrijo, M. H. de S. (2014). *Formação para a cidadania: análise de pesquisas na perspectiva da Educação Matemática Crítica.* Master's thesis. Universidade Federal de Goiás, Goiânia/GO.

Figueiras, L., Healy, L. & Skovsmose, O. (2016). Difference, inclusion and mathematics education: Launching a research agenda. *International Journal for Studies in Mathematics Education, 9*, 15–35.

Freire, P. (2000). *Pedagogia da indignação:* cartas pedagógicas e outros escritos. Unesp.

Gadotti, M. (2010). *Escola cidadã* (13th edition). Cortez.

Gomes, M. P. (2003). Índios: o caminho brasileiro para a cidadania indígena. In J. Pinsky & C. B. Pinsky (Eds), *História da cidadania* (pp. 419–445). Contexto.

Guarinello, N. L. (2003). Cidades-estados na antiguidade clássica. In J. Pinsky & C. B. Pinsky, *História da cidadania* (pp. 29–47). Contexto.

Gutstein, E. (2006). *Reading and writing the world with mathematics: Toward a pedagogy for social justice*. Routledge.

Heywood, A. (2010). *Ideologias políticas: Do liberalismo ao facismo*. Ática.

Orwell, G. (1945). *Animal farm*. Secker and Warburg.

Pais, A. (2012). A critical approach to equity. In O. Skovsmose & B. Greer. (Eds), *Opening the cage: Critique and politics of mathematics education* (pp. 49–92). Sense Publishers.

Penteado, M. G. (2001). Computer-based learning environments: Risks and uncertainties for teachers. *Ways of Knowing 1*(2), 23–35.

Santiago, M. & Akkari, A. (2020). Citizenship, social exclusion and education in Latin America: The case of Brazil. In A. Akkari & M. Kathrine (Eds), *Global citizenship education: Critical and international perspectives*. Springer.

Skovsmose, O. (2001). Landscapes of investigation. *ZDM—Mathematics Education, 33*(4), 123–132. Reprinted as Chapter 1 in O. Skovsmose (2014), *Critique as uncertainty* (pp. 3–20). Information Age Publishing.

Skovsmose, O. (2005). *Travelling through education: Uncertainty, mathematics, responsibility*. Sense Publishers.

Skovsmose, O. (2008) *Desafios da reflexão em educação matemática crítica*. Papirus.

Skovsmose, O. (2020). Banality of mathematical expertise. *ZDM, 52*(6), 1187–1197.

United Nations. (2015). *Transforming our world: The 2030 agenda for sustainable development*. United Nations.

9. About Unfinishedness, Dreams and Landscapes of Investigation

Daniela Alves Soares

This chapter aims to propose that landscapes of investigation are possible learning environments for the establishment of unfinished young students' dreams. Thus, through the lens of certain authors, I present some philosophical ideas of the human being, such as those of unfinishedness and dreaming, as well as foreground and totalisation. Next, I briefly explore aspects of the school tradition evidenced by philosophers who have influenced education as it is today, such as students as "blank sheets", school as the space to develop reason, and mathematics as "true reason". Then, I address technical rationality and critical rationality, presenting some differences and identifying paradigms in which we can recognise mathematics education. Finally, I present landscapes of investigation as a possibility in the face of such studies. Therefore, despite this chapter being characterised as a theoretical essay, I present excerpts from observations, interviews and discussion groups that were part of the data production during my doctorate.

This text is based on my doctoral study, which addresses imaginations and dreams, unfinishedness and infinite ideas. All these concepts materialise in a proposal for mathematics education through new paradigms. I begin this journey with the presentation of ideas from certain authors, based on important concepts related to human nature. In a second step, I reflect on the school tradition evidenced by some philosophers who influenced education as it is today. I also propose a reflection on paradigms in which we can locate mathematics education. At the end, I address possibilities in light of these studies. This chapter is set out as a theoretical essay in which I also use excerpts from the data that were part of my doctoral work, such as interviews, discussion groups and observations. I call the excerpts "scenes". Here, I use pseudonyms for the students' and teachers' names.

Scene 1: About the Human Being, Its Dreams and Unfinishedness

"Who is Laís?", I asked her. She laughed, and then she replied: "Difficult question ... [...] This year, I mainly ask myself this question. Because this year, I'm a little lost. I don't know exactly who I am ... I have some ideas about what I want, but I'm not sure ... So, it's a little difficult, like, to say exactly who I am" (Soares, in progress, my translation).

Laís was 16 when I conducted the interviews for my doctoral study. On that occasion, during the second semester of 2019, I interviewed eight students from a Brazilian public school. I conducted a discussion group with these same students and attended some mathematics lessons in the class. Lais' answer to my question, which can be understood as reticent, insecure and open to imagination, represents one of the main human conditions well: unfinishedness. This condition is usually most evident in adolescence. What does unfinishedness represent? What is its origin? On these questions, I will present some ideas inspired by my previously published text (Soares, 2019).

For Paulo Freire (1983), unfinishedness is the human origin of every search. A Being perceives itself as someone who does not know everything and does not achieve everything; it perceives itself as always in process. It is a Being[1] as a verb, not as a noun.

---

1 When I refer to the noun, I will present this term with the first letter capitalised—"Being", to avoid confusion with the continuous tense of the verb to be—"being".

Because the human being knows itself unfinished, it has hopes and dreams. It has hope because this is part of human nature, and the state of unfinishedness, incompleteness, leads it to embark upon a search movement. Hope can be understood as a flashlight that guides one within this movement. It is like a natural will, which compels one to think, feel and act and, therefore, makes human beings not merely bystanders in history, but protagonists. Due to its intrinsic condition of existence, Freire indicates that hope is the natural form of the human being, and not hopelessness. In the author's words: "I am not, for example, first of all a Being without hope who may or may not later be converted to hope. On the contrary, I am first a Being of hope who, for any number of reasons, may thereafter lose hope" (Freire, 2000, p. 50). If hope is intrinsic to Being, hopelessness, then, is a distortion of that. And with hopelessness, living would be a pure succession of events without any interference from Being; it would be pure determinism.

And the Being dreams. When I write dream, I do not mean the dream projections that we manifest when we are sleeping, although these can be representations of what we dream while we are awake. Still, one characteristic is fundamental, and it is precisely the key point that differentiates one from the other: the dream to which I refer is consciously imagined. Furthermore, it represents the yearnings we have as a social and historical human being, originating from our condition of unfinishedness and the desire to be more.

According to Freire (1983), women and men dream because they want to have experiences of freedom, of humanisation. For the author, dreams have a political perspective and are linked to the historical experiences that Beings live through, to their views of life. Hopes and dreams are answers regarding infinity. The human search is motivated, thus, by the human beings' desire to be more.

In this way, the unfinishedness of the human being generates hopes and dreams, and leads the human being to search in a great kaleidoscope of landscapes. Human unfinishedness generates foregrounds.

The Danish mathematics educator Ole Skovsmose (2014a) uses the term foreground to refer to people's visions of the future, which includes their desires, dreams, hopes, fears, obstacles and frustrations. The experiences from the past—which form one's background—and the way the Being understands and interprets these experiences influence one's landscapes of the future, but do not determine these landscapes. New possibilities can be formed; new dreams can emerge. The unfinishedness of the human being makes it an undetermined Being, with multiple possibilities for the formation of foregrounds.

The ideas of the Franco-Lithuanian philosopher Emmanuel Levinas are also based on the concept of unfinished Being. In fact, for Levinas, it is not possible to understand the Being based only on itself; the formation of what one is (subjectivity) depends not only on the Being, but also on its exterior. There is something beyond us that has influence, that is not under our control and that never ends. In this way, the Being turns to something that it is not, something that overflows, and that is part of the human condition. What motivates this metaphysics, this turning to the other, this transcendence? Levinas answers by saying that it is the idea of the infinite (Levinas, 2007). This idea represents the seed of transcendence that exists within all of us, and that constitutes our incompleteness.

Even so, the socially inserted Being often does not recognise the human condition of eternal incompleteness and goes in search of supposed solid definitions that would contain, for example, what knowledge is, what education is, what God is, how a family should be. This supposed exhaustion of Being, Levinas calls a totalisation.

Levinas states that it is not possible for woman and man to be totalised individuals. Because of this condition, the formation of the Being's subjectivity depends not only on the interior, but also on the exterior, the relationship with the world and with others. There is "a surplus always exterior to the totality, as though the objective totality did not fill out the true measure of Being" (Levinas, 2007, p. 22). So, the movement to exteriority, which is transcendence, is a movement of hope, which turns towards what the Being is not and, therefore, turns towards dreams.

Because it is aware of being unfinished, the Being in process is educated. Education is then part of the human search movement. And what is the school that receives this Being in process like?

Scene 2: The School That Receives That Being: Its Tradition

While explaining how to solve the system

$$\begin{cases} x - y = 3 \\ 2x + 3y = 4 \end{cases}$$

teacher Yael addressed the students using expressions like "any questions, guys?", "did everyone understand?", "right?", and one student or another made a gesture with his head confirming or denying. At the end of the explanation of this exercise, some students emitted the "ahhh" sound, symbolising that they now understood it. For me, this dialogue expresses that the teacher is concerned with the students' learning, but it also reveals that the purpose of the class is that the students understand a truth that is set, and not that they dialogue about it and with it. Thus, the mathematics teacher seems to me to be the one who confirms the knowledge already established (Soares, 2022, my translation).

This excerpt, taken from my field notes during observation of classes by teacher Yael, demonstrates what is normally expected from a traditional school environment: that it is a space created to transmit the knowledge accumulated historically by humanity from the most diverse sciences as historical facts, mathematical algorithms and the laws of nature, as well as a space for reaching the truth.

Some philosophers studied the role of knowledge and school in their time, and much of what they wrote has an influence on us today: this is the case with John Locke and Johann Herbart, representatives of empiricism, and René Descartes, representative of rationalism.

Locke, who lived between the years 1632 and 1704, believed that everything that is learned comes through experience, and that schools are one of the environments where students are exposed to this learning (Locke, 2007; also Aranha, 1996). The experience happens, in some way, through knowledge accumulation, because he believed that students arrive at school as "blank sheets", ready to be written upon.

Johann Friedrich Herbart, a German philosopher who lived between the years 1776 and 1841, believed that learning takes place through the development of reason and also through experience. He wrote about educational activity and highlighted instruction as one of the essential functions of school. Regarding instruction, Herbart does not separate intellectual from moral activity, and the teaching process is for him, in some way, guided by interest (Hilgenheger, 2010). He divides the teaching process into five parts: preparation, presentation, assimilation, generalisation and application (Larroyo, 1974).

Both philosophers, Locke and Herbart, even while considering school as a space for students' experiences (Herbart still associated these experiences with the students' interests), highlight the development of reason, through instruction, as one of the most important roles of this institution. School is the space where knowledge is grounded, and students will develop their reason as much as they are encouraged to in that space. In this perspective, the social context is not considered. It is also important to note that such philosophers were always referring to a school for the elite, and not a common one.

On the other hand, Descartes considered that knowledge must develop reason. This philosopher, born in 1596, believed that knowledge is innate to Being, and therefore all knowledge comes from interiority, and not from external factors (Descartes, 1955). Thus, he believed that the origin of knowledge is good thinking, and sought throughout his life to find the design of human thought. During this search, he was disappointed with the areas of knowledge of literature and humanities, as he found in them to contain more doubt than certainty. While pursuing certainty, he was seduced by mathematics, which (with its demonstrations and solidity) seemed to him to be the most auspicious environment for a safe and true reason. He started to follow a path from doubts to certainties, from the statement "I doubt, therefore I think" to the famous one "I think, therefore I am". Descartes also proposed to distribute complex knowledge in parts, starting from the simplest, using the rational method to reach at the truth (Aranha, 1996).

Much of what is understood nowadays by mathematics and by teaching mathematics is, of course, due to Plato (2004), who conceived mathematical objects as perfect and belonging to the world of ideas. We must also give Descartes his dues—he interpreted mathematics as the most suitable scientific process for the search for truth, through reason.

Indeed, not only these philosophers and educators, but so many others, influenced what we traditionally understand by the terms school and teacher: namely the space for the transmission of accumulated knowledge, and the holder of truths, respectively. In addition, during my doctoral data production, I noticed that having a traditional perspective of mathematical knowledge is far from being an isolated case. Following these reflections, some questions associated with mathematical knowledge at school can be raised: what mathematics teaching has been practised in the face of this type of school? How does this school relate to students who, as mentioned earlier, are unfinished and driven by hope? Does this school create conditions for the development of students' dreams and foregrounds? And can mathematics classes create these conditions?

Scene 3: Technical Rationality, the Exercise Paradigm and Mathematics Education

> School is the place where people are most thinking about passing on information to us, and this information is not necessarily going to be used ... it will be more technical, scientific and accurate information ... [Iara]
> And they insist on preparing us for college entrance exams in the future, but they do not provide support to help us decide how the future is going to be. [Giselle]
> Well, I gave you two lists of exercises, one of them was the one with six problems, we already did some of them together, and you were supposed to have done all the other exercises. Any questions, bring them to the monitoring. ... So we are going to start working on another list now, about linear functions. As I said, I will not do all the exercises here during the class, but I will do at least one of each type, so that you can do the others. [Professor Ariel]
> (Soares, 2022, my translation).

The first two excerpts refer to moments shared in a discussion group. At that time, I asked the students what spaces they saw where it would be possible to talk about dreams, both in school in general and in mathematics classes. It is possible to observe, from the dialogue between Iara and Giselle, how the philosophy of education effectively materialises in class. The students did not find any space for hopes and dreams, because they saw school as a place of closed content, that is often technical and scientific. It is school that is the transmitter of accumulated knowledge, of reason, and that is the place where the argument of truth is put into practice.

Thus, the way students learn mathematics at school reflects a philosophical and ideological perspective, based on historically constructed paradigms. The paradigm that I will highlight now is that of technical rationality.[2]

Rationality is the quality of being based on reason and therefore, in this bias, objectivity is prioritised over subjectivities, through logical and often mechanical actions. In addition, the scholar Henry Giroux points out that objective actions, within a certain rationality, are based on a set of interests that define and limit personal perspectives (Giroux, 1983).

This type of rationality is embedded in the educational process and, with its principles of control and certainty, proposes a model that involves the passive reproduction of knowledge. Thereby, the classroom is established as a conformed environment, in which the teacher has the power of knowledge to be transmitted, and the students are responsible for understanding this knowledge, and sometimes for memorising pre-established formulas and techniques.

In the field of mathematics, the exercise paradigm—a term used by Skovsmose (2008)—refers to the endless lists of technical exercises often disconnected from reality. These exercises are usually formulated by an authority outside the classroom and most of the time the students have the objective of finding the only right answer. In this paradigm, the learning of theory occurs through the exhaustion of technical repetition and, in this sense, it is opposed to an investigative activity. Professor Yael's speech, presented in this section, is an example of work in mathematics within this paradigm.

Related to the dimension of technical rationality and, in some way, to the exercise paradigm, is the concept of knowledge transmission. This concept is based on the idea that teachers and students are expected to be passive consumers or knowledge transmitters, instead of negotiators in the world in which they work and act (Giroux, 1983).

Finally, transmissive teaching is, by definition, totalising teaching. After all, teaching based on these assumptions does not open up spaces for doubts, aspirations, for the understanding of Being as a process, much less for transcendence and the infinite.

---

2   This topic, as well as the one that follows, was inspired by a text previously published by Soares and Civiero (2017).

And how are the students who know they are incomplete and hopeful, when faced with a school based on technical and transmissive rationality? The endless lists of exercises proposed by teachers, most of them decontextualised, can give them the safety of learning the technique, but they do not open up possibilities for the risks of being more, of real and sensitive living.

And how are the students in transformation, who are Beings as a verb, in a school that proposes itself as a transmitter of truths through exercises, mostly with a single answer? They become Beings who give up their infinity and believe in totalising truths. If mathematics education based on this paradigm does not constitute the most suitable environment for these Beings in process, unfinished, what other paradigms and environments can be thought of?

Scene 4: Landscapes of Investigation as a New Paradigm: Possibility of Education for the Unfinished Being That Dreams

I asked the students: "How do you think we could create more spaces for dreams in math classes?" The student Iara said that, among the subjects, mathematics is the one that usually is taught as a pre-established knowledge. And so, her group suggested reducing the competition. Because, for them, competition in class creates a kind of pressure ... and "it is very individualistic", the student Giselle said. Therefore, they proposed the group work, thinking that it would help both in the didactic and in the psychological sense, as it is possible to debate a problem and talk about it. Giselle added that sometimes someone cannot solve one problem, but can solve another, and that is why it is possible to "share ideas". This group argued that when one person explains to another it is much easier to understand the problem because it becomes more personal. Isabela added: "This thing of discussing a problem together, joining ideas', with that, a less hostile environment would be created, and people would have more affinity with mathematics", Giselle concluded (Soares, 2022, my translation).

In the second topic, I highlighted some philosophers who have influenced what we now understand as the role of the school. One of them was Locke, who despite valuing the school as a space for experience, perceived students as blank slates for knowledge. Another was Herbart, who valued control and discipline as moral practices within school and also valued instruction, with its greater objective of reason development. Even though he highlighted the importance of developing students' interest, Herbart's method was not based on the value of subjectivities and social and cultural practices. Descartes, on the other hand, was a philosopher who accredited great importance to reason, arguing that it is possible to submit all kinds of knowledge to the order of reason, and make mathematics one's greatest exponent, as the most exemplary representative of the search for truth.

In the third topic, I highlighted how these philosophies of education can be related to the model of technical rationality, which is also connected, in mathematics classes, to the exercise paradigm. As I highlighted, in this paradigm mathematical knowledge is acquired by students in a generally mechanical and repetitive way. This means that in this paradigm, one avoids privileging personal interpretations, and most exercises are not based on reality, much less on foregrounds. Thus, this model reveals a feature of totalisation of knowledge, from the perspective of control, which does not create space for unfinishedness, critical thinking, hopes or dreams.

In contrast, Skovsmose (2001) proposes a paradigm that is based on the so-called critical rationality, which places less importance on technique and more on reflection, especially socio-political reflection.

He presents learning environments that represent this new model of rationality, and that value the process of collective exploration and investigative search. In these learning environments, collaboration among students is prioritised, and it is expected that all communication takes place through dialogue. In these environments, called landscapes of investigation (Skovsmose, 2001), since students accept a challenge to be investigated, they become co-responsible for the learning process. My conversations with Iara, Giselle and Isabela revealed that, in the students' opinions, working in groups to solve problems would be interesting for mathematics classes. In the investigative environment, learning is a process, and therefore not static. The teacher creates the scenario and proposes the challenge, and from that moment on they become an adjunct in the situation to be investigated. Thus, students are considered protagonists of the process, their personal knowledge and experiences are valued, their background is valued; they are not blank slates! Faced with such a student role, the teacher has no control over the process; the teacher does not know what will happen. Therefore, this process involves risk but also numerous possibilities. In Skovsmose's words:

> A landscape of investigation invites students to formulate questions and to look for explanations. The invitation is symbolized by the teacher's "What if...?" The students' acceptance of the invitation is symbolized by their "Yes, what if...?" In this way the students become involved in a process of exploration. The teacher's "Why is it that...?" provides a challenge, and the students' "Yes, why is it that...?" illustrates that they are facing the challenge and that they are searching for explanations (Skovsmose, 2014b, p. 7).

The figure presented in Skovsmose (2014b), also presented in Chapter 1 of this book, reveals two important movements: one horizontal, and the other vertical. The horizontal is the passage from the odd environments (1), (3) and (5) to the even ones (2), (4) and (6). This passage highlights the priority change, from technique to criticism (landscapes of investigation). This means a passage from control to openness, from single answer to multiple possibilities, from "solve this" to "how would it be". It is, therefore, a movement of transcendence, also comparable to the idea of infinity, already presented in this text.

The vertical movement, on the other hand, reveals an openness in relation to the topics covered in class, starting from pure mathematics (environments (1) and (2): totally abstract and closer to the "ideal" world), passing through semi-reality (environments (3) and (4): intermediaries) and reaching reality (environments (5) and (6): more concrete and within the real world). This opening reveals possibilities for less platonic mathematics, less associated with the great Cartesian truths, which is free to follow paths based on context, culture and subjectivities. These are learning environments that provide more space for students' lives, for their desires, for their problems, as well as for their interests, horizons and dreams. Environments (5) and (6), therefore, assume an opening for the observation and construction of foregrounds.

In landscapes of investigation, mathematics is not a closed subject, but it is open to the possibilities of the world and of history. I think this is what the students wanted to highlight in the conversations with me: we need to make mathematics more human, and accordingly, less competitive. After all, students can find multiple answers, and therefore different truths, that are not definitive. In this sense, with just one more question—"What if ...?", or "Why?", or "Is it fair?"—new possibilities can arise. Beyond that, the challenge proposed by the teacher is also unfinished and incomplete during a landscape of investigation. And that is why frustrations and hopes, imaginations and dreams can happen.

To summarise, when investigating the conditions that lead the human being to embark upon the search process, I focussed on concepts such as unfinishedness, foregrounds, dreams and transcendence. This study made me question what the school that receives the searching Being is like. In that sense, the traditional school has some problematic aspects: the school tradition seems to go in the opposite direction to that of a school open to dreams. It shows itself as a totaliser and transmitter of knowledge and, in terms of mathematics classes, it appears to be a technicist, closed to the exteriority. In an extremely unequal society, as is still the case in Brazil and in many developing countries, hopes and dreams represent very important features for personal and political development. And the school has a fundamental role in this, especially with the adolescent public. In this sense, I understand that it is necessary to have schools, and especially mathematics classes, committed to political and social aspects. After all, reducing mathematics learning to technique means restricting students' foregrounds and dreams.

Thereby, I conclude that landscapes of investigation represent possible learning environments that are more coherent with what is most human among all of us. In other words, I highlighted in this chapter the concepts of incompleteness, and the desire to be more (from Freire); the idea of the infinite and the transcendence (from Levinas), and the development of future horizons, as foregrounds (from Skovsmose). After all, Laís, Iara, Giselle, Isabela and so many other students may like mathematics exercises, but they do not dream about them.

Acknowledgements

I would like to thank all my graduate colleagues from *Épura*, who have always helped me with important considerations about my studies; also, to the teacher, the students and the school who so kindly welcomed my data production; and finally, to DAAD (*Deutscher Akademischer Austauschdienst*) for the scholarship.

References

Aranha, A. M. L. de (1996). *Filosofia da educação*. Moderna.

Descartes, R. (1955). *The principles of philosophy*. Dover Publications.

Freire, P. (1983). *Educação e mudança*. Paz e Terra.

Freire, P. (2000). *Pedagogy of freedom: Ethics, democracy, and civil courage*. Rowman and Littlefield Publishers.

Giroux, H. A. (1983). *Theory and resistance in education: A pedagogy for the opposition*. Bergin and Garvey Publishers.

Hilgenheger, N. (2010). *Johann Herbart*. Massangana.

Larroyo, F. (1974). *História geral da pedagogia*. Mestre Jou.

Levinas, E. (2007). *Totality and infinity: An essay on exteriority*. Duquesne University.

Locke, J. (2007). *Some thoughts concerning education: Including of the conduct of the understanding*. Dover Publications.

Plato. (2004). *Republic*. Hackett Publishing Company.

Skovsmose, O. (2001). Landscapes of investigation. *ZDM—Mathematics Education, 33*(4), 123–132.

Skovsmose, O. (2014a). *Foregrounds: Opaque stories about learning*. Sense Publishers.

Skovsmose, O. (2014b). *Critique as uncertainty*. Information Age Publishing.

Soares, D. A. (2019). Sobre sonhos e empoderamento de jovens: possibilidades para as aulas de matemática. *XXIII Encontro Brasileiro de Estudantes de Pós-Graduação em Educação Matemática (XXIII Ebrapem)*. São Paulo: Universidade Cruzeiro do Sul.

Soares, D. A. (2022). *Sonhos de adolescentes e reflexões sobre as aulas de matemática*. Doctoral dissertation. Universidade Estadual Paulista (UNESP), Rio Claro, Brazil.

Soares, D. A. & Civiero, P. A. G. (2017). Problematizando a racionalidade técnica por meio dos ambientes de aprendizagem da educação matemática crítica. *VIII Congreso Iberoamericano de Educación Matemática (VIII Cibem)*. Madrid: Universidad Complutense de Madrid.

10. A Dialogue in Eternity: Children, Mathematics, and Landscapes of Investigation

Ana Carolina Faustino

> This chapter aims to discuss the connection between landscapes of investigation and dialogue shown through interactions amongst children in mathematics classes in the early years of elementary school. The characters Socrates, Lakatos, Galileo, Skovsmose and Carolina meet in eternity to comment on the relationships between landscapes of investigation and dialogue in mathematics education. Thus, this chapter is a fictitious dialogue in which the reader will encounter a dialogical text where Carolina will introduce some examples on children's interaction in a landscape of investigation. The dialogue involves children from Grade 3 and Grade 5 mathematics classes in a public school in the state of São Paulo, Brazil.[1]

I need to hurry. Wow! How late I am! I look hastily at my watch, hold my laptop tighter and prepare to run. I go up higher and higher. The route is nice, but it is uphill. Uphill and uphill and uphill. There are beautiful views. Several times, I stop to enjoy the view and also to catch my breath. There are many flowers along the route. Flowers of all colours surround the route towards Eternity. Near Eternity, I am going to meet Socrates,

---

[1] These classes were audio-recorded and notes were taken in the researcher's field diary. The recordings were listened to several times, which led to the identification of critical events that were transcribed and analysed.

Galileo, and Lakatos. They have all promised to meet me to talk about dialogue. I am very grateful for their readiness to do so. I forwarded my PhD dissertation to them in advance, and I wonder if they have found time to take a look at it. I am also anxious, as they are all so famous.

Socrates: O Carolina, are you alright? You look flushed! You must be tired. Do you fancy a glass of wine?

Carolina: Hello, Socrates! Hello, Lakatos! Hello, Galileo!

Galileo: Hello!

Lakatos: Hello, Carolina!

Carolina: Wine? Yes, I would like some. Thank you, Socrates.

Socrates: Here you are. By the gods, Carolina, would you mind telling me what that shiny object next to you—similar to a rectangle—is?

Carolina: This is a machine, a laptop computer; I will use it so that Ole Skovsmose can take part in our conversation and talk with us a little.

Galileo: Where is he at this very moment? Is he squeezed into the machine?

Carolina: No! What we will see is just his image on a screen. Right now, he is at the university where he works, where there is a machine just like this one. By using it, he will be able to see us and hear us, and we will also be able to hear him and see his image.

Socrates: Dear Zeus! How interesting!

Carolina: I am connected! The machine is working just fine! Hello, Ole! How are you? Socrates, Lakatos, Galileo, and I are waiting for you to get started. Can you see and hear us?

Skovsmose: I am fine and I can hear you! This meeting makes a lot of sense to me! Let us get started!

Socrates: Is that your advisor?

Carolina: Yes, yes. Socrates, Galileo, and Lakatos, this is my advisor, Ole Skovsmose. He is also dedicated to studying critical mathematics education, which is concerned with dialogue and its relationship to mathematical learning, democracy and critique.

Lakatos: It is a pleasure to meet someone who is also dedicated to mathematics!

Carolina: The connection seems to be quite good.

Galileo: What connection?

Lakatos: Wait, Socrates wants to say something.

Socrates: Carolina, my pricey friend, I am most grateful for your invitation to converse about children, dialogue and landscapes of investigation. I believe that it will be an invaluable opportunity to all of us. Plato granted me the opportunity to address various issues in his dialogues. In *Banquet*, I was able to exalt love; in *Meno*, I addressed the possibility of teaching virtue, or not; in *Republic*, I addressed justice, education and the forms of government; in *Laques*, we talked about courage; other relevant issues were discussed in other dialogues. Nevertheless, I have never had the opportunity to talk specifically about dialogue. Until now. I am not very well acquainted with the subject, but I am fond of the possibility of examining this topic with thee and being able to learn about landscapes of investigation. I am very happy with this invitation indeed.

Carolina: It is an honour for me to talk and learn with all of you. Could you briefly explain why dialogue is important to you?

Socrates: To think is to dialogue with our own soul.

Carolina: So, thinking can be understood as a dialogue! An inner dialogue. Great! What about the dialogue in which we share our conceptions and our way of understanding the world with others?

Socrates: Certainly. The first advantage of exposing our ideas out loud during a dialogue is the possibility of convincing ourselves of what we speak about, of clarifying the arguments we use for ourselves, of examining possible contradictions in our own discourse. However, the advantages of dialogue do not end here. The inquiries can help to make the other person clarify some point that they did not understand; they can still present refutations. Here, we have the second advantage of dialogue. Through them, the interlocutor can make their arguments known, being able to convince the one who hears them. Refutations and inquiries allow interlocutors to refine their arguments and seek the truth.

Lakatos: I agree!

Galileo: Me too.

Socrates: Dialogue is a path in which the pace and direction are adjusted with each step. Being aware of the advantages of a dialogue, should I choose or not choose this form of interaction to seek knowledge in the field of philosophy?

Skovsmose: Yes! This makes sense.

Carolina: Did you charge your interlocutors for the dialogues?

Socrates: Certainly not. I am not a sophist like Protagoras, Gorgias or Hippias!

Skovsmose: You talked to poor and rich men. In Plato's dialogue, entitled *Meno*, one of his interlocutors was Meno's slave.

Socrates: Yes, Skovsmose. For me, in that dialogue, the goods that a man possessed were not important, nor whether the man devoted himself to philosophy, to calculation. I wanted to interrogate someone and, for that, it was only necessary that we could communicate. What did Meno's slave and I need to have in common in order to communicate?

Carolina: To communicate ... it is important that two human beings speak the same language.

Socrates: Certainly. It was only important that Meno's slave spoke the same language as me: Greek. So I was able to ask him some questions, such as: "So doubling the side has given us not a double but a fourfold figure?"[2]

Carolina: It is important that everyone speaks the same language in order to communicate. In addition, I am surprised that you all speak English so well, Socrates. The Eternity really did you good.

Skovsmose: Meno is a very famous dialogue between mathematical educators because of the passage in which you use the interrogative method with the slave, and the slave learns geometry.[3]

Carolina: Yes! My dearest friends Galileo and Lakatos, how do you like the opportunity to talk about dialogue?

Galileo: Well, in view of my writings, I think I can contribute. In the book *Dialogue Concerning the Two Chief World Systems*, I also employed dialogue to discuss different worldviews upon the geocentric and heliocentric systems.

Lakatos: I agree! After coming to Eternity, my PhD dissertation was made into the book *Proofs and Refutations: The Logic of Mathematical Discovery*. I feel I will also be able to contribute to our discussion. In the book, I created an imaginary dialogue. I would very much like to know what landscapes of investigation look like.

Skovsmose: Landscapes of investigation are learning environments that support investigative work in which students are actively involved

---

2   Here, Socrates demonstrates having an eidetic memory. In this passage, he uses the same words as in Plato's book (Plato, 1956, p. 133), Of course, this memory must have to do with the oral tradition of which Socrates is a part. Thus, all of his work is in an oral format, an orality that Plato sought to portray in his dialogues.

3   For a critical analysis of the dialogue between Socrates and Menon's slave, see Fernandez (1994).

in the teaching and learning process. They are learning environments that provide spaces for investigations. They are inclusive and support investigation. They also open up space for dialogue.

Lakatos: I am fond of the word *investigation*.

Galileo: Would you care to explain?

Skovsmose: Yes, of course. *Highlighting the feature of investigation serves the purpose of putting aside some of the many ways the notion of dialogue can be used, such as: 'The dialogue between Russia and the USA includes tensions' and 'The dialogue in the film Godfather is well written'. By highlighting 'investigation', the notion of dialogue becomes related more directly to an educational context.*[4] A landscape of investigation is accessible to all students. It is a learning environment that provides learning opportunities for all students in the classroom, regardless of who they are. The researcher Amanda Moura,[5] for instance, has been dedicated to investigating landscapes of investigation that create learning opportunities for both hearing and deaf students—that is, learning environments that are accessible to all students who are present in a certain classroom.

Lakatos: Excellent! I see! I see!

Socrates: How fortunate I am, Skovsmose. Galileo asked about the word *investigation* and you presented the meaning of this word, and specified some of the characteristics of landscapes of investigation. Now that dialogue is a distinctive feature of a landscape of investigation, can you try to clarify why this feature is important?

Skovsmose: Landscapes of investigation also make room for dialogue. During the investigation process, students play an active role in their learning. This means that they need to make decisions. In order to accomplish that, they need to share their perspectives, present and defend their arguments, and suggest a perspective to be followed by the group. The qualities of dialogue support criticism in mathematics classes.

Socrates: In other words, landscapes of investigation enable all the individuals involved in the process to engage in dialogue in mathematics class. Is dialogue identical in mathematics classes amongst children, young students and adults?

---

4 See Faustino and Skovsmose (2020, p. 9).
5 See Moura (2020).

Skovsmose: Carolina, I would like you to talk a little about how dialogue can emerge in a landscape of investigation in mathematics classrooms in the early years of elementary school. Could any sort of learning environment favour the emergence of dialogue?

Carolina: For the past few years, I have been researching the process of interaction between teachers and children, and amongst children themselves, in mathematics classrooms. My aim is to understand how elementary school teachers and children could bring dialogue into their classrooms.

Socrates: Well then, if you have been pursuing this goal for the last few years, you may have come to the truth about what dialogue is. Would you agree?

Carolina: Well, I do not know for sure whether I have found out the truth about what dialogue is. However, I feel that I have made important observations about dialogue in the early years of elementary school. I have discussed how children and teachers engage in this process, and how the communication patterns that emerge in the classroom can influence how mathematics learning takes place. I started by visiting a school and observing mathematics classes in the early years of elementary school. I did so to learn from teachers and children how interaction takes place in these classes. During one semester, I followed the development of the "Environment and Mathematics Project". This was organised by two teachers: one who taught Grade 3 students, and one who taught Grade 5 students. That experience of going to the two classrooms enabled me to understand how those teachers and students interacted and put dialogue into action.

Socrates: O Carolina, what is the age range of the students at that teaching level?

Carolina: They are between six and ten years old on average.

Socrates: Most of my interlocutors in the dialogues were male young adults.

Galileo: So were mine! Simplicio, Salviati and Sagredo were male adult characters created by me to represent the different views between the Copernican and the Ptolemaic systems, which were dominant in the society I lived in at that time.

Lakatos: In the dialogues I created, a teacher converses with his students who are called Alpha, Beta, Gamma, Delta, Epsilon, etc. These characters were all fictional, the fruit of my imagination. Nonetheless, I like the idea that real students who populate classrooms can engage in dialogues about their mathematical thoughts.

Skovsmose: Paulo Freire also studied the concept of dialogue, but with adult peasants. On the other hand, Helle Alrø[6] and I approached this concept based on the interaction amongst teenagers in mathematics classes in schools in Denmark.

Lakatos: How compelling!

Carolina: I problematise the concept of dialogue by taking children into account based on their interactions during mathematics classes. The data collected are evidence supporting the fact that the dialogic process can occur in a classroom with children. Teachers and children can establish dialogic interactions to learn mathematics.

Galileo: Excellent!

Lakatos: Why is it important for children to learn mathematics through dialogue?

Carolina: Education is political.[7] So is the way teachers interact with their students in the classroom. Learning to engage in dialogue and learning mathematics in a dialogic way are actions connected with the promotion and maintenance of the democratic principles in our society. When children learn mathematics through dialogue, they also learn to respect others' perspectives, which are different from their own. They learn that other students' perspectives can be a resource for learning. Children will learn how to listen actively, how to develop arguments to justify their perspectives, and how to share the mathematical strategies they are using. Children also learn to reconsider their own perspectives and think critically about mathematical knowledge, thus developing a critical attitude towards that knowledge. Children learn to work in groups and help each other, to engage and collaboratively solve the proposed activities. Through dialogic interactions, children will learn mathematics and also how to interact democratically. Dialogue can also

---

6   Alrø and Skovsmose (2004).

7   The political dimension of education has been emphasised both in the field of education (Freire, 1994) and in the field of mathematical education (Skovsmose and Valero, 1999; Skovsmose, 2007, 2011, 2014; and Skovsmose and Greer, 2012). Freire demonstrates an uncertainty regarding the purposes of education, as it can contribute to the domestication of students or their liberation. The uncertainty regarding the purposes of mathematics education is also present in Skovsmose's theorisations (2011, 2014). The author points out that mathematics education can empower or disempower students.

give room for children to perceive themselves as human beings who produce culture and knowledge.

Lakatos: That connection between dialogue and democracy is certainly fascinating. I appreciate this idea. I would like to know though how children aged six to ten years old can learn to engage in dialogue.

Carolina: The classroom functions as a microsociety in which children and teachers with different backgrounds meet for the purpose of learning. The learning of dialogue takes place in the process of engaging in dialogue itself. Teachers and children learn how to interact in a more dialogic way every day, and end up humanised.

Socrates: It just occurred to me... You claim that children and teachers can converse. I was wondering if you could elaborate upon how you reached that conclusion. What does dialogue mean when interacting with children?

Carolina: During interactions between children and teachers studying mathematics in the early years of elementary school, it was possible to identify two patterns of communication: the "sandwich" pattern of communication, and dialogue.[8]

Socrates: Dear Carolina, what do you mean by the "sandwich" pattern of communication? I have never heard the word "sandwich" before.

Lakatos: A sandwich is something one can eat!

Socrates: Ah...

Carolina: The "sandwich" pattern of communication is characterised by the presence of the teacher's question, then the student's response to that question, followed by the teacher's assessment, which is conveyed through feedback. In this type of interaction, teachers have a centralising and predominant role. Students, though, have minimal responsibility for their learning processes, as they only need the teacher's explanation for the answers provided. By interacting through this pattern, teachers and students do not create a favourable environment for the development of critical reflections on mathematical knowledge. Instead, the only worldview to be shared is the teacher's, which must be learned and repeated by the students. This pattern of communication does not favour the construction of mathematical knowledge from students' backgrounds and experience.

---

8   See Alrø and Skovsmose (2004). In Mehan (1979) this pattern is also called initiation-response-evaluation (IRE).

Galileo: Carolina, would you care to provide us with an example?

Carolina: Yes, I will. Here is an example from my PhD dissertation:

Teacher: What do we have here? [pointing at the drawing on the board]

Students: A semi-straight line. [answered in unison]

Teacher: An AB semi-straight line, isn't it? How do I know that this is a semi-straight line and not a straight line or a straight-line segment?[9]

Galileo: I see. There is the teacher's question, the students' response and the teacher's assessment.

Carolina: Exactly! Note that the teacher's question is a closed-ended one and it requires a specific answer that she already has in mind. Thus, it is up to the students to guess what the answer is. In that excerpt, it is also possible to notice that the teacher does not use terms like 'right' and 'wrong' when giving feedback. Instead, she repeats what the students said when their answer is correct. This demonstrates that the teacher was trying to interact openly with them, but was using a communication pattern based on the teacher being the one who shares knowledge, and the student, on the other hand, being the one who must acquire that knowledge without their background and experience being taken into account.

Socrates: I apologise, Carolina. I did not pay attention to the very last part. I was serving some more wine to Lakatos. Would you mind repeating what you were saying?

Carolina: The "sandwich" pattern of communication does not contribute to the teaching and learning process by taking the students' knowledge into account. This pattern of communication has as its starting point the teacher's view only, which must be learned and repeated by the students.

Lakatos: What were the conditions in which that communication pattern emerged?

Carolina: At the end of the semester, the Grade 5 teacher let me know that she was falling behind schedule with the mathematics content and, although she knew we were supposed to work on the project that day, she said she would rather work on the angles content, and I agreed. The students were not grouped that day, but sitting in rows. The teacher

---

9   Faustino (2018, p. 129), my translation.

stood at the front of the room, drew something on the blackboard, and interacted with the children based on the explanation she had already given them the day before. She then asked a question and one or more students answered. Following this, she gave them feedback. Right after that, she handed out the textbooks and the students worked on angles exercises. When the teacher interacted with her students through the "sandwich" pattern of communication, she had the impression that she was talking the students through the mathematical content, step by step. This communication pattern is a result of systematisation and pressure to fulfil the curriculum. Whether the students had learned that content or not, the teacher explained it, thus she did her job. The students then work on the course book exercises, correction follows, and the teacher is free to move on to the next topic. The data presented in this part of the PhD dissertation suggest that the way the teacher lays the classroom out influences the communication pattern that will emerge. Planning her class around the textbook, having students sitting in rows, and centralising interaction in the role of the teacher fosters the emergence of the "sandwich" pattern of communication.

Lakatos: Well, Carolina, I see you've forgotten about your glass of wine. Let's make a toast, shall we? Cheers!

Skovsmose: Cheers!

Carolina: Cheers!

Socrates: Cheers!

Galileo: Cheers! I most appreciate your arguments, Carolina. It is fundamental to me that one is able to reflect critically on previously acquired knowledge rather than just replicate that.

Socrates: Carolina, thou ought to see that, in the "sandwich" pattern of communication, teachers only transmit their knowledge to students, thereby missing the great opportunity to get to know what the latter know. In my dialogues, first and foremost I tried to listen, I tried to understand what my interlocutors knew.

Carolina: Agreed!!! Listening actively to children is a fundamental aspect of dialogue. It is essential that teachers, who possess a great deal of knowledge on the topic covered, use that knowledge to actively listen to children and find a way to interact with them in order to help them to reflect critically on the object of knowledge. During active listening, teachers undergo an epistemological shift in order to understand the student's worldview. The necessary conditions are created so that children's knowledge is the starting point of the teaching and learning process. Active listening[10] and the epistemological displacement that is required in order to understand students do not take place in a vertical relationship, in which the teacher knows everything and the child knows nothing.

Socrates: Indeed! I knew that I knew little and, therefore, during my wanderings in the Agora, I sought to engage in dialogues to learn more and understand what others knew.

Carolina: Exactly! In 2017, I went to the Mathematics Education and Society conference in Greece and visited the architectural ensemble of Acropolis and Ancient Agora. It is an absolutely inspiring place. And recognising our incompleteness makes us move towards knowledge. Paulo Freire highlighted that the dialogue process only takes place when the vertical relationship between teachers and students is overcome.[11] The teacher is no longer the one who just teaches, and the student is no longer the one who just learns. The teacher starts to learn while teaching, and the student to teach while learning. Active listening, as well as the teacher's displacement to understand the student's worldview, can contribute to overcoming the vertical relationship between teachers and students.

Socrates: Carolina, the type of interaction that presents the characteristics of the "sandwich" pattern of communication differs from dialogue. In that case, you have two essential questions to answer. First of all, what is dialogue? Secondly, in which context did dialogue take place?

Carolina: It will be a pleasure to share what I mean by dialogue with you! Dialogue happens when participants in the teaching and learning process share their worldviews about the object of knowledge—in this case, mathematics. To understand students' worldviews, the teacher seeks to build a horizontal interaction, in which they listen and move epistemologically. In a dialogue, different students share different views of the world, and new views of the world regarding the object of knowledge can emerge from that exchange. The construction of new worldviews is the result of a collective discussion, and it takes shape in students' learning.

---

10   Alrø and Skovsmose (2004).
11   Freire (2014).

Galileo: I most appreciate this understanding of dialogue as sharing and building new worldviews. It is essential that the human being get acquainted with different views on the object of knowledge. Even though most people believed in the geocentric system, in which the sun revolves around the earth, there were others—like Nicolaus Copernicus and I—who believed in a heliocentric system, in which the earth revolves around the sun. The existence of different worldviews, and the sharing of those views, is extremely important for the construction of knowledge. Carolina, with my experiences in mind, I would like to share a concern with you.

Carolina: Go ahead. Each and every concern is welcome.

Galileo: When I presented my ideas about the heliocentric system, which were contrary to most people's worldviews at that time, my work *Dialogue Concerning the Two Chief World Systems* was prohibited. I used arguments that were based on mathematical calculations and astronomical observations through a telescope to defend my ideas about the earth revolving around the sun. However, arguments of power—which only had strength because they came from people who represented authority at that time—were used to discredit the epistemological arguments that I was presenting. According to the authorities, my worldview should not be shared, but repressed. In order to refute my worldview, the Inquisition[12] condemned me. Intolerance, as always, governed the whole process. I fought dogmatism, but in order to survive I had to totally deny my worldview. In this sense, I wonder how teachers and students can share their worldviews in the classroom.

Carolina: Galileo, I have to say this is an excellent question and it helps me to continue explaining what I mean by dialogue. In order for different worldviews to be shared between teachers and children who seek to learn collaboratively, three characteristics are essential: maintaining equity, taking risks and making an investigation.[13] The first one, maintaining equity, concerns the fact that, although teachers

---

12 The Roman Inquisition was founded in 1552 by Pope Paul III and aimed to ensure that the canons of the Catholic Church were followed. To that end, the Roman Inquisition condemned books, and people who presented arguments that were contrary to the Church's position. In 1600, Giordano Bruno, who was also inspired by the Copernican system, was burned alive under the order of the Roman Inquisition. To more about the role of the Roman Inquisition regarding the development of the Copernican system, see Mariconda (2020).

13 Alrø and Skovsmose (2004).

and children are different, they can have a dialogue in a horizontal relationship as long as the strength of their arguments is not based on the more or less powerful position that they hold in the classroom, but on the arguments that are used instead.[14] Both children and teachers can present relevant and coherent arguments to justify their worldviews. It is important that this relationship based on equity is promoted in the classroom, and teachers have a key role in promoting it. That can be read in my thesis, in the short story "Hand in Hand With Mathematics and the Environment", in which the Grade 5 teacher tried to create conditions for all students to express themselves before she presented her point of view. At the same time, the teacher tried to interact with them by encouraging the ones who had not yet presented their perspectives to do so. Thereby, during the interaction process, teachers can help children who have not had the opportunity to express themselves. Teachers can also interact with children by making sure that everyone has the opportunity to share their worldviews and present their arguments.

Galileo: I could not agree more, Carolina. The strength of an argument is in its coherence. Would you care to tell me what you mean by doing investigations and taking risks?

Carolina: Conducting an investigation means to engage in a collaborative search process, to move towards acquiring more knowledge. The investigation process cannot be imposed, and it takes shape through the teacher's invitation for students to enter a scenario for investigation. When students accept the teacher's invitation, they are guided by their curiosity, which is expressed in the construction of perspectives that are designed based on the object of knowledge and then shared. Taking risks refers to the unpredictability of the different worldviews that might emerge in the classroom, the varied paths that dialogue can take and the different responses that might arise during the investigation process. In this sense, teachers create an environment in which they prepare the investigation with a purpose, but are open to the different directions that the dialogue between them and students can take. During the investigative process, a student can take a path that had not been anticipated by the teacher. In this sense, risks are taken. However, taking risks also means unpredictability and new possibilities:

---

14 Skovsmose (2007).

possibilities for new worldviews to be built from the dialogue between children and teachers.

Galileo: Have you found any specificity in the dialogue amongst children regarding sharing and building new worldviews?

Carolina: A specific feature of the dialogue amongst groups of children is the choice of perspective that the group will follow. This way, after sharing their perspectives and defending them by using mathematical arguments, it is necessary to decide which one will represent the group. It could be a new perspective that emerged from those presented in the group, or it could be one presented by a student individually. Early years students are learning to have a dialogue and choose the perspective that will represent the group, and that part of the process can be a challenge for them at times. In the data presented in the short story "Estimating Lengths", the different groups created different strategies to help them select the perspective to be followed. In one of the groups, it was possible to note that children are able to associate each of the shared perspectives with the student who presented it, and that choosing one of them to represent the group might be interpreted as though there is only one winner. In this sense, we consider it important that, when interacting with children, the teacher makes sure they understand that the selected perspective is the result of group work and that all the others presented in the group are the result of collaborative work, which means that nobody wins or loses.

Galileo: Interesting! Indeed, it might be difficult to choose between different perspectives, but this is a very rich process. I wonder how the choice was made in the other group.

Carolina: Let me tell you. To select the perspective, the other group decided to vote on the estimated height of the giraffe. Although voting is a practice related to democratic principles, we believe that it enabled the children to continue with the dialogue. It is a way of solving the problem of choosing perspectives. It is important that early years teachers and students discuss the ways in which they are selected in each group, and also that the teacher interacts with students by helping them to build the chosen one based on the coherence of the proposed arguments. Such an interaction allows children to learn how to choose the perspective in a more dialogic way, as well as to associate it with a collaborative process and collective construction, thus disconnecting their minds from the individual and competitive aspect of "winning" and "losing".

Socrates: Carolina, my honest friend, I most appreciate what you have been highlighting upon the concept of dialogue. My first query was answered. However, the second still remains. Would you mind telling me how you identified a dialogue amongst the children?

Carolina: Not at all. In addition to the characteristics I mentioned earlier, Alrø and Skovsmose[15] highlight that some speech acts with special characteristics can be identified during dialogue. Such acts were called dialogic acts, as follows: getting in contact, locating, identifying, advocating, thinking aloud, reformulating, challenging and evaluating. The presence of a variety of these acts in an interaction between teachers and students in the classroom allows the interaction to be classified as dialogic.

Lakatos: From what you said, however, Alrø and Skovsmose[16] identified the presence of those acts in mathematics classes, which were aimed at teenagers and not children. Is that right?

Carolina: Precisely! After having observed the interaction between teachers and students for a semester, it was possible to identify the presence of a variety of dialogic acts in the interaction between teachers and Grade 3 children. The dialogic acts were there.

Socrates: Dear Carolina, what would each of the dialogic acts you mentioned earlier be? Could you use some examples to elucidate them?

Lakatos: The examples of dialogic acts both please me and make me uneasy at the same time. Carolina, have you read my work *The Logic of Mathematical Discovery: Proofs and Refutations*?

Carolina: Yes, I have.

Lakatos: Were you able to identify dialogic acts in the interactions amongst characters in my work? Perhaps you could mention some excerpts and exemplify some of the dialogic acts.

Carolina: Allow me to read an excerpt from your work, Lakatos. Let's have a look. After the teacher discussed the proposed activity and shared their perspective on it, the students interacted as follows:

> *Pupil Alfa*: I wonder. I see that this experiment can be performed for a cube or for a tetrahedron, but how am I to know that it can be performed for *any* polyhedron? For instance, are you sure, Sir, that *any polyhedron, after having a face removed, can be stretched flat on the blackboard*? I am dubious about your first step.

---

15   Alrø and Skovsmose (2004).
16   Alrø and Skovsmose (2004).

Pupil Beta: Are you sure that in triangulating the map one will always get a new face for any new edge? I am dubious about your second step.[17]

In this excerpt, we can notice the presence of two acts: recognising and thinking out loud. The students Alpha and Beta recognised the teacher's perspective by examining it carefully, and then they put the dialogic act into action: thinking aloud. They say out loud what their doubts are regarding the teacher's perspective, and ask questions that make room for the teacher to justify it.

Lakatos: Point taken.

Carolina: In another excerpt, we have the following interaction:

Lambda: But let me come back to the *second possibility*: when we do *not* find any local counterexample to the suspect lemmas.
Sigma: That is, when refutations do not assist proof-analysis! What would happen then?[18]

In this excerpt, the student Lambda highlights a point that he believes they should discuss; he takes a stance. Taking a stance means creating a line of argument that demonstrates the way the student is approaching the task and, in this case, drawing other students' attention to a specific aspect. Then, the student Sigma demonstrates getting in contact and reformulating when he says: "That is, when refutations do not assist proof-analysis! Getting in contact is a dialogic act and is expressed by the preparation for dialogue and by paying attention to others, which can be emphasised by the complementation of phrases. When Sigma says "That is", he demonstrates that he was actively listening to the other student and getting in contact. In other words, he was paying attention to what the other student was saying. Then, Sigma says: "[...] when refutations do not assist proof-analysis!" and reformulates what Lambda had just said. Therefore, he repeats in a slightly different way what the other student had said. This repetition allows Sigma to explain what he understands by using different words, and additionally demonstrates that both share the same perspective. By asking the question "What would happen then?", Sigma makes room for the other student to present justifications and for new mathematical ideas to be recognised.

---

17   Lakatos (1976, p. 9).
18   Lakatos (1976, p. 51).

Lakatos: I am very pleased to be acquainted with the fact that the dialogic form used by me in the writing process includes dialogic acts. Carolina, enlighten me please, will you? How do these acts appear in dialogues amongst children and amongst real students of your research?

Galileo: I am also interested in getting to know how those acts contributed to the students' learning.

Carolina: Numerous dialogic acts were identified in the interactions between teachers and children, and within their groups. Getting in contact is a preparation for dialogue in such a way that children pay attention to each other, look at each other and start to be silent, showing that they are ready to engage in dialogue. For children, this act is very important, since it gets everyone ready to talk about the object of mathematical knowledge, and they gradually learn that the subject to be addressed concerns investigation. Topics that do not concern the investigation are abandoned during the act of getting in contact. What moves them now is a curiosity about the learning object, which will be shared in the dialogue. The act of getting in contact has great prominence at the beginning of the dialogue, but at the same time, it permeates the whole process. It is essential for maintaining the dialogue and can emerge, for example, when a student complements another's speech, demonstrating that they are listening actively. Getting in contact also emerges when a student uses tag questions to end their sentences, thus creating conditions for another participant to continue the dialogue. In the short story "Filming the World", the children used tag questions to seek support for their perspectives from other children or the teacher.

Socrates: Very proper. I have one more query; you said that to create a dialogue it is necessary to have several dialogical acts. You explained one of them. What other dialogic act was identified?

Carolina: The act of reformulating, for instance. It allows the teacher to use the formal mathematical language and help children to take it in. By repeating what children say in a different way, the teacher can establish connections between the terms that the child is using, which come from their knowledge or experience, with the terms of mathematical language. Thereby, the connections between the terms are highlighted, and their meanings are made explicit in the context in which they are being used. In this sense, it is possible for the children to understand the connections between their way of expressing themselves and the mathematical language, which enables them to incorporate the mathematical language into their own vocabulary. Such an act also creates opportunities for the teacher or the children to check if they understand what others really meant.

Lakatos: In which context did dialogue emerge?

Carolina: Like I said, during the interaction between teachers and children, and within groups of children, we were able to identify several dialogic acts, which enabled the identification of dialogic interactions. However, we also identified some acts that do not contribute to promoting dialogue. Instead, they close it or degenerate it into other communication patterns. These acts are called *non-dialogic acts*.

Socrates: Dear Carolina, would you be kind enough to exemplify such non-dialogic acts?

Carolina: Of course. Please, have a look at the data presented in the tale "Filming the World: How Much Water Do I Use Daily?"[19]

Julia: If every time someone flushes the toilet they spend six litres of water each time.
Carla: If a person goes to the toilet like [pause] let's say three times.
Julia: Sometimes I go to the toilet five times a day. [Everybody laughs, including Julia: Then Julia stops.]
Julia: Oh! That does not matter!
[Carla suggests they should make a decision, so they could move on.]
Carla: Three times, for example. Let us consider three times as being a day's average [pause].
Denis: Is it three times eight?
[Clearly enough, the number of times a person flushes the toilet has to be multiplied by the number of litres used in a flush.]
Julia: Yeah, but [pause]
Carla: No. Why do we multiply three by eight?
Julia: If we have six litres of water for each flush?
Denis: Six litres of water [pause]
Julia: Six litres.
Denis: On each toilet flush.
[While Denis talks, he waves his hands and points to Julia, who nods. Apparently, Denis has remembered incorrectly that the amount of water in a flush is eight rather than six litres. This correction being made, some calculations can be completed.]
Julia: A person goes to the toilet three times. So, how are you going to do it?
Denis: Three times six.
Julia: So why did you say eight?

---

19   Faustino (2018).

Denis: Because I thought she said eight litres.
Julia: So, you are going to calculate that [pause] three times six is eighteen.
Carla: The result is eighteen litres.
[Carla turns and looks at Pedro, who has been quiet during the whole conversation.]
Carla: Eighteen times six. Pedro, it is six times, isn't it? Are there six people in your house?
Julia: He does not even know what he is talking about! [laughing and putting her hand to her mouth. Pedro does not say a word. With his hand holding the pencil in front of his mouth, he nods, confirming that there are six people in his family.]
Carla: So, six times eighteen, which is?[20]

Lakatos: Carolina! In this excerpt, I noticed the presence of the act of recognising.

Carolina: Exactly! Carla shares her reasoning with the phrase "three times, for example. Let us consider three times as being a day's average", and Denis recognises in Carla's perspective a mathematical algorithm, which he complements with "three times eight?".

Lakatos: I really like seeing children dialoguing about mathematics.

Carolina: Absolutely! During the dialogue, some of the children composed their interaction based on dialogic and non-dialogic acts. For example, by asking a question, Carla invites Pedro to talk about the number of people in her house. Julia, on the other hand, does not wait for the student to respond and interacts in a way that disqualifies him when she says: "He does not even know what he is talking about!". This non-dialogic act is named *disqualifying* and involves making negative comments about other students, which emphasise their lack of skills to contribute with the dialogue due to their lack of knowledge. This act can hinder children's engagement in dialogue and the construction of a positive cultural identity, as well as impacting negatively on the way they see themselves as human beings capable (or not) of producing mathematical knowledge. Dialogic and non-dialogic acts become an important tool for teachers and researchers to analyse the interaction between students in the mathematics classroom, and to identify whether the interactions have some elements of power, even the ones which are dialogic. By identifying non-dialogic acts, teachers can help students to reflect on their actions and gradually learn to interact based on respect and equity.

---

20   Faustino (2018, pp. 150–151), my translation.

Lakatos: Carolina, how does your research contribute to the field of study? What does it bring to the table?

Carolina: The data from my research provide evidence of the presence of dialogic acts amongst children during mathematics classes in the early years of elementary school. Children are conceived of, therefore, as dialogic beings that produce knowledge in the relationships they establish with the world and with others. When teachers set the environment in order to favour the emergence of dialogue, they value children's mathematical knowledge. As a result, classes are no longer adult/teacher-centred and shift to being focussed on children's mathematical knowledge, on their unusual way of relating to the object of knowledge. That does not mean that teachers will abandon all explanations; rather, they will start with students' mathematical knowledge and, by moving epistemologically to understand that, they will be able to create rich situations of sharing and negotiating meanings with the whole class. Throughout this process, teachers can find aspects that can be expanded and systematised, which allow them to work with both the knowledge that children bring to school and with academic knowledge. In this sense, the moment the teacher allows the construction of a concept based on their students' mathematical knowledge and realises that the explanation of a concept or generalisation can be carried out, they will do it without objectifying the student. That is certain since the vertical logic between teachers and children is broken when they start from the children's mathematical views.

Lakatos: For me it is a great idea to start to understand children's mathematical strategies.

Skovsmose: For me it also makes sense and could lead to new ways to approach a problem during the investigation. Please go on, Carolina.

Carolina: Therefore, the dialogic mathematics class is based on children's mathematical knowledge and the sharing of their mathematical worldviews. In the dialogue, the child learns mathematics through the relationships they establish in the classroom and, during the dialogue, they also learn to respect and value diversity. Thereby, dialogue can help children learn mathematics and develop a democratic posture in the classroom. Dialogue also creates possibilities for children to interpret the social context in which they are inserted, and begin to act towards transforming it. The research also enabled the identification of both the eight dialogic acts and the eight non-dialogic acts, which constitute an important tool for researchers, children and teachers to use to co-build a dialogic mathematics class.

Socrates: Zeus! We have run out of wine!

Carolina: Well, I have the impression that I have been able to elaborate on everything I wanted to say so far and also to reflect on the inquiries made by the three of you. Possibly, tomorrow I will have new ideas to discuss and considerations to share, but as far as today is concerned, I am content, and I leave you with a big smile on my face for having this dialogue.

Socrates: My dearest Carolina, allow me one last comment. I regard the interaction we established in this meeting as permeated by dialogic acts. Would you agree with that?

Carolina: Definitely, Socrates! Thank you very much! Our interaction has been extremely rich in dialogic acts. We get in contact when we sit at the gates of Eternity paying attention to each other and engaging in dialogue to learn about a specific theme. We identify different approaches to the process of having dialogues with young and adult men in Ancient Greece, with both women and men, mainly peasants, at the end of the 20th century, as well as dialogues with children at present times. I think out loud by making public what I've been learning about dialogue. Galileo challenged me with an excellent question about relations of power and dialogue. Moreover, our dialogue turned out to be an interaction that changed me. Thank you to each and every one of you!

Socrates: I can assure you that the three of us are very glad indeed of this dialogue with you and Skovsmose...

Lakatos: ... Do come back to eternity.

Galileo: You are most welcome! Sharing worldviews with you has been a delight!

Acknowledgements

I would like to thank Amanda Queiroz Moura, Celia Regina Roncato, Daniela Alves Soares, Debora Vieira de Souza, Denner Dias Barros, João Luiz Muzinatti, Miriam Godoy Penteado, Ole Skovsmose, Roger Miarka, Gustavo Barbosa, Patricia Linardi, Raquel Milani and Jackeline Rodrigues for their helpful comments and suggestions.

References

Alrø, H. & Skovsmose, O. (2004). *Dialogue and learning in mathematics education: Intention, reflection, critique*. Kluwer Academic Publishers.

Faustino, A. C. (2018). *Como você chegou a esse resultado?: O diálogo nas aulas de Matemática dos anos iniciais do Ensino Fundamental*. PhD dissertation. Universidade Estadual Paulista (Unesp). Available at: http://hdl.handle.net/11449/180358

Faustino, A. C. & Skovsmose, O. (2020). Dialogic and non-dialogic acts in learning mathematics. *For the Learning of Mathematics, 40*(1), 9–14.

Fernandez, E. (1994). A kinder, gentler Socrates: Conveying new images of mathematics dialogue. *For the Learning of Mathematics, 14*(3).

Freire, P. (2014). *Pedagogia do Oprimido* (58th edition). Paz e Terra.

Lakatos, I. (1976). *Proofs and refutations: The logic of mathematical discovery*. Edited by John Worral and Elie Zahar. Cambridge University Press.

Mariconda, P. R. (2000). O diálogo de Galileo e sua condenação. *Cad. Hist. Fil. Ci., Campinas, Série 3, 10*(1), 77–160.

Mehan, H. (1979). *Learning lessons*. Harvard University Press.

Milani, R. (2015). *O processo de aprender a dialogar por futuros professores de matemática com seus alunos no estágio supervisionado*. PhD dissertation. Universidade Estadual Paulista (Unesp). Available at: http://hdl.handle.net/11449/124074

Moura, A. Q. (2020). *O encontro entre surdos e ouvintes em cenários para investigação: das incertezas às possibilidades nas aulas de Matemática*. PhD dissertation. Universidade Estadual Paulista (Unesp), Rio Claro, Brazil.

Plato (1956). *Protagoras and Meno* (W. K. C. Guthrie, Trans.). Penguin Group.

Skovsmose, O. (2007). *Educação Crítica: Incerteza, Matemática, Responsabilidade*. Tradução: Maria Aparecida Viggiani Bicudo. Cortez.

Skovsmose, O. (2011). *An invitation to critical mathematics education*. Sense Publisher.

Skovsmose, O. (2014). *Critique as uncertainty*. Information Age Publishing.

Skovsmose, O. (2014). *Foregrounds: Opaque stories about learning*. Sense Publisher.

Skovsmose, O. (2014). *Um convite à educação matemática crítica* (Translation Orlando Andrade Figueiredo). Papirus.

Skovsmose, O. & Greer, B. (2012). *Opening the cage*. Sense Publishers.

Skovsmose, O. & Valero, P. (2001). Breaking political neutrality: The critical engagement of mathematics education with democracy. In B. Atweh, H. Forgasz & B. Nebres, B. (Eds), *Sociocultural research on mathematics education* (pp. 37–55). Lawrence Erlbaum.

11. Inclusive Landscapes of Investigation

Ole Skovsmose

> By means of two examples, one concerning polygons and one concerning erosions of democracy, I characterise the conception of inclusive landscapes of investigation. These are teaching-learning environments that are accessible for everybody, and invite dialogue across differences. This brings me to refer to universal design, which provides a broader perspective on the construction of inclusive environments. Finally, I relate the concept of critique to the characteristics of inclusive landscapes of investigation.

Inclusive mathematics education tries to provide learning environments for all groups of students. Inspired by the work of the Épura research group, the idea that inclusive mathematics education could provide learning environments evolved: environments where all students, independent of particular differences, can *learn together*.[1] This leads me to the idea of forming inclusive landscapes of investigation.

In literature one can find two different interpretations of inclusive mathematics education, which I refer to as the specific and the general. According to the specific interpretation, inclusive mathematics education concerns students with disabilities such as, for instance, blind or deaf students. The book *Inclusive Mathematics Education: Research Results from Brazil and Germany*, edited by David Kollosche et al. (2019), addresses

---

[1] The Épura research group was founded in 2008, and is associated to Unesp in Rio Claro. Épura members are, first of all, Master's and PhD students and researchers working with inclusive education inspired by critical mathematics education. The group is coordinated by Miriam Godoy Penteado and Ole Skovsmose.

problems related to this interpretation. According to the general interpretation, inclusive mathematics education concerns students learning together across cultural, economic and political differences, as well as across differences with respect to learning capacities. The book *Diversity in Mathematics Education: Towards Inclusive Practices*, edited by Bishop et al. (2015), addresses this general interpretation. In the following, I have both the specific and the general interpretation in mind.

In this chapter I discuss the notion of *inclusive landscapes of investigation*,[2] which are landscapes intended to be accessible for different groups of students, whatever the differences concerning their abilities or social diversities. When the notion of landscapes of investigation was first developed, I did not have inclusive mathematics education in mind. However, now I want to extend the discussion of landscapes by incorporating concerns about inclusion.

I am going to present two examples, which will serve as references for the following discussion: the first, *Polygons*, relates to the specific interpretation of inclusive mathematics education, while the second, *Erosion of Democracy*, relates to the general interpretation. After the examples, I will outline a general characteristic of inclusive landscapes of investigation. As a conclusion, I will address the notion of critique, and in this way relate the discussion to the concerns of critical mathematics education.

1. Polygons

The landscape *Polygons* was developed in an inclusive setting in Brazil, where deaf and hearing students were learning together. The example is described by Amanda Moura (2020) and by Amanda Moura and Miriam Godoy Penteado (2019).[3]

In Brazil it has become common practice to integrate children with different diagnoses of disability into the regular school system and not to let them remain in specialised institutions. In the city of Rio Claro in the São Paulo State, one finds "inclusive schools" that receive students

---

2   For a general presentation of landscapes of investigation, see Chapter 1, "Entering Landscapes of Investigation".
3   See also Chapter 12 by Moura and Penteado.

with disabilities. Here one finds classrooms with, for instance, both deaf and hearing students, as was the case in the classroom where Moura conducted her study.

One investigation in which the students (around ten years old) were engaged concerned the classification of geometric figures. The students were presented with a huge number of figures cut out of cardboard and were asked to group the figures according to characteristics they found relevant. Some figures had curved edges, some had straight edges and some simply looked strange. How could they sort things out?

This activity lead to the question: What is a polygon? Another question quickly emerged: What sign in Libras should they use for the word "polygon"? (Libras is the Brazilian sign language used by deaf people). During the process, deaf and hearing students worked together in groups, and in order to facilitate the communication, an interpreter who could speak Libras was around. However, there was also direct communication between hearing and deaf students. For years the students had been in the same class, and the hearing students had learned some Libras.

One possibility in sign language is to do a spelling out of the letters, making them with the hands: P-O-L-Y-G-O-N. However, such spelling is a last resort, as spelling out too many words makes communication in Libras slow and awkward. It is more efficient to try to decide upon a single sign for the word "polygon". Libras is a language in construction, and many concepts do not have a particular sign. With respect to mathematics, there is no well-defined extension of the regular sign system, and nothing called "mathematics in Libras". This makes it relevant to negotiate signs for particular concepts, such as "polygon". But which sign should we use?

The discussion led to some mathematical clarifications of the notion of polygon. One is reminded of the process described by Lakatos (1976), where the sequence of proofs and refutations leads to further clarifications of the notion for polyhedron. In a similar way, the complexity of the notion of polygon was revealed through the discussion about which sign to use for this concept. One problem in choosing a particular sign was trying not to assume that one is dealing with a particular polygon, like a rectangle for instance. The sign should refer to the general properties of a polygon, and not to a particular group of polygons. In the end, the students decided upon a sign, namely the sign for "many" followed by the sign for "lines", repeated with the hands moving in different directions showing a "P".

Polygons turned into an inclusive landscape of investigation where the participation of both deaf and hearing students was not only possible, but necessary. The students had to judge the adequacy of the suggested signs, considering what signs already exist in Libras as well as the significant mathematical properties of a polygon. An important feature of an inclusive landscape of investigation is that different groups of students can come to work together. An inclusive landscape of investigations facilitates meetings across differences, as was the case with the landscape *Polygons*.

2. Erosions of Democracy

The Weimar Republic was formed in 1919 and destroyed in 1933, when Adolf Hitler came to power. During that period a disastrous erosion of democracy took place. One could consider to what extent similar erosions have taken place elsewhere in the world. I find that a broader discussion of democracy is vitally important today, where non-democratic movements and authoritarian discourses seem to be gaining more and more influence. I see this as being a challenge also to mathematics education. As an illustration, let us consider *Erosions of Democracy* as a possible landscape of investigation, and to what extent it can become inclusive, considering the general interpretation of inclusive education.

This landscape is a thought experiment, which as far as I am aware has not been tried out previously. However, I have presented this thought experiment on different occasions, such as at the *Segundo Colóquio de Pesquisa em Educação Matemática Crítica* (Second Colloquium in Research in Critical Mathematics Education), Rio Claro, 2018, and at *the Primeiro Encontro Mato-Grossense de Professores que Ensinam Matemática* (First Meeting for Mathematics Teachers in Mato Grosso), Tangará da Serra, 2018. After these presentations I received many comments and suggestions which inspired me to carry out a further elaboration of the thought experiment.

Previously (Skovsmose, 1994), I have highlighted that democracy refers to at least four sets of ideals that concern: procedures for voting; fair distribution of welfare; equal opportunities and obligations; and rights to express oneself. Each of these ideals can be eroded. In many discussions about democracy, the existence of procedures for voting have been considered as definitional for a democracy. My point in mentioning the four sets of ideals is that a democracy only gets established through a variety of components, the right to vote being only one of them.

Voting is an act through which one allows somebody else to speak in one's name. The investigation of voting procedures is a mathematical issue, which can be challenging also for university students in mathematics (see Steiner, 1988; and Obraztsova and Elkind, 2012). However, the landscape *Erosions of Democracy* can focus on some specific issues such as the problem of "a tyranny of the majority". As an illustration of what this could mean, one can consider a small community of seven people, four coming from the north and three from the south. Every decision in this community is based on voting. The four from the north vote together. It has to be decided who is going to do the manual work. With four votes against three, it is decided that people from the south have to do it. Who is going to pay taxes? With four votes against three, it is decided that people from the south will have to pay. In the end, it is voted that the three people from the south should serve as slaves in the small community. The possibility of a tyranny of the majority shows that while voting constitutes part of democratic procedures, it far from ensures democracy. But how are we to eliminate the possibility of a "tyranny of the majority"? This question may provide an entrance to further investigations of procedures for voting.

One can imagine functioning democracies in both rich and poor countries. Much more difficult to imagine is a functioning democracy in a country with extreme differences between rich and poor. Decisions in a democracy are expressions of a shared will, and if extreme differences are maintained, one must suspect that the shared will has become subsumed by particular interests. There are different ways of mathematically describing distributions of welfare; one is referred to as the Gini Coefficient. However, a more elementary mathematical technique can also illustrate a distribution of welfare. As part of *Erosions of Democracy*, the students can investigate the distribution of welfare in imagined countries with a population of, say, 100 people each. The distributions of welfare in such imagined mini-countries can be the same as the distributions in real countries, and in this way the students can explore principal features of economic inequalities the world over.[4]

In a democracy, one assumes equity with respect to the right to vote, but equity relates to any relationship with the law. For instance, the students can investigate to what extent differences with respect to juridical procedures can be related to a people's political position, economic situation or ethnicity. A specific issue concerns the way in which the police and the penitentiary system are operating. With reference to Brazil, statistics show that in 2019 the police in Rio de Janeiro killed 1810 people.[5] One can follow up and clarify the ethnic distribution behind such a number. One can also consider the number of people put in jail in Brazil, or in any country for that matter, and see to what extent one can identify any ethical biases. Where it is possible to identify marked biases, one might question to what extent everybody is treated equitably with respect to the law. This is an important issue with respect to *Erosions of Democracy*.

The classic issue of freedom of speech concerns possibilities for expressing oneself, as well as the right to articulate opinions about political, religious, cultural or any issues for that matter. In a democracy, the general obligation of the press is to give a voice to everybody, and the students could consider to what extent such rights are observed or ignored by the various media outlets. There are different ways of addressing such an issue, and a starting point can be taken from the way Reginaldo Britto investigates the visibility of white and black children in different magazines (see his chapter "Media and Racism" in this book). The same investigative procedure can be used for revealing different degrees of visibility of various politicians in different media outlets depending on their ethnicity, gender or political position. An investigation of visibility can show to what extent a voice is given to everybody in an equal way, or if certain media outlets operate with particular biases.

In my imagination, *Erosions of Democracy* may become inclusive by the way it calls for contributions from students with different experiences

---

4 Inspiration for such an approach can be found in Smith (2011).
5 See *BBC News: Rio Violence: Police Killings Reach Record High in 2019*. https://www.bbc.com/news/world-latin-america-51220364

and backgrounds. In Denmark—reflecting a general European trend—one finds a growing animosity towards groups of people who are considered immigrants and are described as being "foreigners". This trend causes an erosion of democracy—for instance, with respect to equity in a variety of situations. In order to address such erosion, it is important to bring together students with different experiences and backgrounds. I see such interactions between those labelled as "foreign students" and those claimed to be "not-foreign students" as an important resource for addressing erosions of democracy. Exploring the landscape *Erosions of Democracy* does not presuppose any homogeneity of the students. Rather, the exploration of the landscapes benefits from diversities. In this sense, I think of it as an inclusive landscape of investigation.

3. Features of Inclusive Landscapes of Investigation

Like any landscape of investigation, inclusive landscapes provide space for investigations. Such landscapes are not organised by sequences of problems to be solved, or by exercises to be answered. Rather, they provide invitations for students to engage in inquiry processes. Questions can be raised, and answers can be suggested, leading to new questions.

The discussion of *Polygons* and *Erosions of Democracy* brings me to highlight the following two features of *inclusive landscapes of investigation*:

1. Inclusive landscapes of investigation provide learning milieus that are *accessible to everybody*. Differences among students do not cause specific conditions for entering and moving around in such landscapes. One might meet a variety of challenges that acknowledge diversities amongst students. The very conceptions of students being "normal" or "not-normal", "foreigner" or "not-foreigner", having "abilities" or "disabilities" lose significance in an inclusive learning environment.

2. Inclusive landscapes of investigation *invite dialogues across differences*. Establishing conditions for dialogue is a general feature of a landscape of investigation. In inclusive landscapes, such dialogue is also supposed to take place across differences, whatever kind of differences one might have in mind. There

could be differences with respect to cultural backgrounds, religious convictions, nationalities or economic conditions, as well as with respect to abilities.[6]

"Inclusive landscapes of investigation" is far from being a well-defined label, and I am not trying to provide the notion with more specificity than what was just outlined. However, I want to condense this specificity into one statement: *inclusive landscapes of investigation invite meetings amongst and across differences.*

The notion of *universal design* was coined by Ronald Mace, who was an architect preoccupied with designing environments accessible for everybody, independent of their physical conditions. In 1963, Selwyn Goldsmith (1997) published *Designing for the Disabled*, which highlighted the idea of ensuring the free movement of all, including people with disabilities, such as people in wheelchairs. It was such a concern that Mace captured with the expression "universal design".

One can think of inclusive landscapes of investigation as an example of universal design (see Marcelly, 2015, for a presentation of this idea). This is an important comparison, although with some limitations. With reference to architecture, "moving around" is a rather well-defined physical property, while the possibility of "moving around" in a landscape of investigation is not a simple quality of the landscape as such. It is also a quality of the way the landscapes are acted out in educational practice.

In the case of *Polygons*, the activities could have become non-inclusive if, for instance, the group organisation of the students had been different by allocating hearing students to one group and deaf students to another. Naturally, being allocated to the same group is not a sufficient condition for establishing inclusive education. It is, for instance, also important to create conditions for the students to communicate with each other and to explore issues together. *Erosions of Democracy* could become non-inclusive if the investigations were differentiated according to the mathematical abilities of the students. It could transform into non-inclusivity as well if political, economic or cultural differences were to define the groupings of the students. Again, more conditions

---

6   A similar characteristic can also be found in Skovsmose (2019). Here one also finds further references to people who have developed the notion, such as Roncato (2015).

are necessary for establishing inclusive learning environments, such as developing a shared interest in acknowledging and understanding different worldviews.

4. Dialogue across Differences

I see critical activities as being rooted in dialogues.[7] This is the reason that critical mathematics education holds a particular interest in creating landscapes of investigation, which provide conditions for establishing dialogue between students and teachers, and amongst students themselves.

The concern for creating inclusive landscapes of investigation is in line with this overall idea. However, through inclusive landscapes, one provides conditions for establishing dialogues *across differences*. Such dialogues are important extra resources for critical activities. I will illustrate this point by again referring to *Polygons* and *Erosions of Democracy*.

Critique has an epistemic as well as a socio-political dimension.[8] The epistemic dimension is illustrated by the landscape *Polygons*, while *Erosion of Democracy* illustrates the possibility of developing a socio-political critique.

The landscape *Polygons* was accessible for both hearing and deaf students; in fact, the presence of both groups of students was crucial for conducting a critical epidemic investigation of the notion of polygon. When one considers the properties of a polygon that could define the sign in Libras, it is important that the perspectives of both hearing and deaf students are articulated, and that everybody engages in a dialogue where ideas are not just stated, but discussed, dissected and developed.[9] A dialogue across differences is an important resource for critical reflections of an epistemic nature.

The landscape *Erosions of Democracy* is inclusive to the extent that everybody is invited to contribute to its exploration. In fact, the presence of a diversity of perspectives is important for addressing questions like: To what extent is systemic poverty an obstruction for a functioning

---

7 For an elaboration of this point, see Chapter 1.
8 See my discussion of critique in Chapter 1.
9 For a further discussion of how negotiations of sign contribute to the shared learning of deaf and hearing students, see Sales, Penteado and Moura (2015).

democracy? To what extent are different groups of people treated differently with respect to legal procedures? Can such differences be related to people's political opinions? To their economic status? To their ethnicity? Which specific examples can one refer to? For addressing such questions and other issues concerning possible erosions of democracy, dialogues across differences are essential. It is important that different voices are not only heard, but also engaged in a dialogue. With respect to the thought experiment *Erosion of Democracy*, we can only speculate about the interaction among the students, but in order to critically address democratic issues, I find it is important to establish dialogues between different worldviews.

I find that dialogues across differences provide important extra resources for critical activities of both epistemic and socio-political natures. For this reason, it is important to explore possibilities for establishing inclusive landscapes of investigation. It is also my hope that such critical activities might support acknowledgements of diversities without falling into dominant discursive stereotypes.

Acknowledgements

I want to thank Denner Barros, Ana Carolina Faustino, Peter Gates, Amanda Queiroz Moura, João Luiz Muzinatti, Miriam Godoy Penteado, Célia Roncato, Daniela Alves Soares and Débora Vieira de Souza for their helpful comments and suggestions.

References

Biotto Filho, D. (2015). *Quem não sonhou em ser um jogador de futebol? Trabalho com projetos para reelaborar foregrounds*. Doctoral dissertation. Rio Claro, SP: Universidade Estadual Paulista (Unesp).

Bishop, A. Tan, H. & Barkatsas, T. N. (Eds) (2015). *Diversity in mathematics education: Towards inclusive practices*. Springer.

Faustino, A. C. (2018). *Como você chegou a esse resultado? O processo de dialogar nas aulas de matemática dos anos iniciais do Ensino Fundamental*. Doctoral dissertation. Rio Claro, SP: Universidade Estadual Paulista (Unesp).

Goldsmith, S. (1997). *Designing for the disabled*. Architectural Press.

Kollosche, D., Marcone, R., Knigge, M., Penteado, M. G. & Skovsmose, O. (Eds) (2019). *Inclusive mathematics education: State-of-the-art research from Brazil and Germany*. Springer.

Lakatos, I. (1976). *Proofs and refutations*. Cambridge University Press.

Marcelly, L. (2015). *Do improviso às possibilidades de ensino: Um estudo de caso de uma professora de matemática com estudantes cegos*. Doctoral dissertation. Rio Claro, SP: Universidade Estadual Paulista (Unesp).

Moura, A. Q. (2020). *Cenários para investigação e escola inclusiva: Possibilidades de diálogo entre surdos e ouvintes em aulas de matemática*. Doctoral dissertation. Rio Claro, SP: Universidade Estadual Paulista (Unesp).

Moura, A. Q. & Penteado, M. G. (2019). The role of the interpreter of Brazilian sign language in the dialogue among deaf and hearing students in mathematics classes. In M. Knigge, D. Kollosche, R. Marcone, M. G. Penteado & O. Skovsmose (Eds), *Inclusive mathematics education: Research results from Brazil and Germany* (pp. 253–270). Springer.

Obraztsova, S. & Elkind, E. (2012). Optimal manipulation of voting rules. In Conitzer, Winikoff, Padgham & van der Hoek (Eds), *Proceedings of the 11th international conference on autonomous agents and multiagent systems* (AAMAS 2012).

Roncato, C. R. (2015). *Cenários investigativos de aprendizagem matemática: Atividades para a autonomia de um aprendiz com múltipla deficiência sensorial*. Master's thesis. São Paulo, SP: Universidade Anhanguera de São Paulo.

Sales, E. R., Penteado, M. G. & Moura, A. Q. (2015). A Negociação de Sinais em Libras como Possibilidade de Ensino e de Aprendizagem de Geometria. *Bolema, 29*(53), 1268–1286.

Skovsmose, O. (1994). *Towards a philosophy of critical mathematics education*. Kluwer Academic Publishers.

Skovsmose, O. (2011). *An invitation to critical mathematics education*. Sense Publishers.

Skovsmose, O. (2019). Inclusions, meetings, landscapes. In D. Kollosche, R. Marcone, M Knigge, M. G. Penteado & O. Skovsmose (Eds), *Inclusive mathematics education: Research results from Brazil and Germany* (pp. 71–84). Springer.

Smith, D. J. (2011). *If the world were a village: A book about the world's people*. Second edition. Kids Can Press.

Steiner, H. G. (1988). Mathematization of voting systems: Some classroom experiences. *International Journal of Mathematical Education in Science and Technology, 19*(2), 199–213.

12. Meetings amongst Deaf and Hearing Students in the Mathematics Classroom

Amanda Queiroz Moura and
Miriam Godoy Penteado

> In this chapter, we are aiming to contribute to practices in mathematics classes that value the idea of meetings amongst differences. We present landscapes that were designed specifically in order to favour interaction among deaf and hearing students. The discussion is based on data produced in a public elementary school in Brazil, in a fifth-year class with both deaf and hearing students. The results highlight that the organisation of tasks based on dialogic communication is essential in order to respond to unpredictability and to enable collaboration in conditions of equity in the mathematics classroom.

Inclusive education now forms part of a global movement aimed at fostering equity. It presupposes a student-centred education that fully serves their educational needs Inclusive education can be seen as a movement for the universalisation of schooling and quality education.

We adopt a perspective of inclusive education as one that supports the belief that those people belonging to under-represented groups have the same educational opportunities as most of the population. It is a perspective based on equity and social justice, aimed at benefiting all students, mainly those who have been historically excluded, but today have gained the legal right to education. Advocating these rights implies recognising that this does not depend only on benefits through legislation. Undoubtedly, having appropriated laws is necessary, but it is also necessary to consider that there are oppressive and exclusionary pedagogical practices that present themselves as barriers to the learning of many students.

This is a challenge that has been faced by researchers from different fields, particularly in mathematics education. We understand inclusive mathematics education as a perspective that is concerned with breaking down discriminatory views regarding the mathematics learning potential of all students. This perspective implies the redefinition of the objectives of schooling, as well as the development of mechanisms that enable the modification of existing school structures and practices, which have been based on classification, segregation and exclusion (Figueiras, Healy and Skovsmose, 2016).

In classroom meetings that embrace pupil differences, a challenge is to create learning environments in which all students—regardless of their differences—can learn together. For Moura (2020, p. 190, our translation), "thinking of inclusive education in terms of meetings amongst different people is to think of an education that practices tolerance". This idea is based on the recognition and appreciation of the differences. It is an education that sees possibilities in learning together.

This chapter presents a discussion based on the results of a research study that aimed to understand the interactions in a classroom where deaf and hearing students meet to study mathematics, through a teaching proposal based on landscapes of investigation. Aimed at contributing to practices in mathematics classrooms that value the idea of meetings amongst differences, we present landscapes that were organised particularly to favour interaction among deaf and hearing students.

The research data was produced in a public elementary school located on the outskirts of a city of the state of São Paulo, Brazil. At the school, there are students in different conditions, and Libras (the Brazilian sign language) is used by both deaf and hearing people. Our research put forward a pedagogical proposal based on the bilingual perspective of education. It was a PhD study where the first author of this chapter was a PhD student and the second one was her supervisor.

The research participants were three teachers, three interpreters of Libras and seventeen students in the fifth grade, aged between nine and twelve years old. Twelve of the students were hearing and five were deaf. These students had studied together since their first year in school and had since been learning Libras, including the hearing students.

The data production involved a group of students with teachers and sign language interpreters studying theoretical aspects of landscapes of investigation, with an emphasis on dialogic communication. With this perspective, the group of teachers planned lessons that took into account the curriculum and the diversity of students in the classrooms. Some of these lessons were recorded in audio and video, and these recordings, in addition to the researcher's field notes, constituted the data for analysis.

In what follows, we present some theoretical connections between dialogue and landscapes of investigation. Then we present details of the constitution of the landscapes by teachers and sign language interpreters. Finally, we present reflections on the interaction among deaf and hearing students during the tasks, and how the organisation of that contributes to the construction of equity and inclusion.

1. Landscape of Investigation and Dialogue

The proposal of a landscape of investigation favours ways of communication that can challenge classes constrained by the exercise paradigm. One of its main features is the possibility of dialogic interaction between participants, as it allows students to participate more actively in the teaching and learning processes. They are a contrast to conventional classes, in which there are no spaces for questioning and/or justifications about the relevance of the tasks, where the value is only on whether the answer is correct or not (Skovsmose, 2001).

Assuming that the quality of communication in mathematics lessons influences the quality of learning, Alrø and Skovsmose (2004) understand dialogue as a typical pattern of communication in a landscape of investigation. This means that dialogue is understood as a conversation aimed at learning. This form of interaction allows perspectives to be shared and discussed between teachers and students. When considering dialogic communication in the context of an investigation, Alrø and Skovsmose highlight its potential to build a condition of equity among participants. For them, equity in an investigation refers to the ways of dealing with the difference. This is related both to emotional aspects and to the way in which the content of the dialogue is treated. Building and maintaining this equity is an ongoing task.

An investigation is guided by curiosity. Dialogical communication helps us to understand the problem being investigated and to build new perspectives. This also includes collectivism and collaboration. For dialogic communication to take place, it is necessary for students to be involved through the acceptance of an invitation and for a task not to be imposed upon them. The lines of investigation take shape from the exploration of students' perspectives, which should not take place as a form of transmission, but as part of a collaboration. It is necessary to make room for new insights to visualise a problem or its solution through new perspectives.

Due to its unpredictability, this approach involves risks. There are uncertainties when one wants to know what the other thinks. Even if one can imagine what could happen, there is no guarantee of what the answer will in fact be. Coping with these risks can also be related to the students' emotional engagement. For example, they may feel uncomfortable during the investigative process, having an opinion contested or rejected by others. On the other hand, they may feel joyful when the shared perspective helps the investigation. On the other hand, unpredictability in this context can also mean new possibilities for learning. It contributes to the students' autonomy during the process.

There is no way to guarantee the nature of the learning, considering that the dialogue is unpredictable and surrounded by uncertainties. But knowing what the other thinks is already a big step towards creating new meanings. The teaching and learning processes are complex due to diverse factors, some of which are the communicative patterns that are enacted. It is through communication that a landscape is created, and it is possible to implement agreed actions in a cooperative way. Cooperation favours equity in the attempt to understand what the other says. In this way, dialogue becomes fundamental when it has to do with inclusive practices (Milani, 2015).

Based on these ideas, Skovsmose (2019) highlights the close connection between dialogic communication and inclusive education. From the dialogue, what also arises is the notion of meeting amongst differences which advocates for an inclusive education perspective that seeks going across differences, avoiding categorising people with disabilities against some standard of normality. Based on this perspective, in what follows, we present how landscapes were built, keeping in mind the encounters amongst deaf and hearing people in mathematics class.

2. Building Landscapes of Investigation

The teachers and interpreters started by forming a study group to learn about theoretical and practical aspects related to the notion of landscapes of investigation. Their discussions focussed on promoting reflections on the perspective of dialogue, on the possible learning environments of a landscape of investigation as proposed by Skovsmose (2001), as well as on the teaching and learning of mathematics in the context of an inclusive school. The group planned tasks for investigation focussing on possible interactions between deaf and hearing students.

The fifth-year teacher chose to develop a task as part of a multidisciplinary project on a healthy diet and the cost of living. The project involved the students going to a supermarket to buy ingredients for a fruit salad. This generated a discussion on the amounts and calories of fruits, their benefits and the costs of a balanced diet. In addition to making decisions when buying fruit, the children were invited to investigate some mathematical properties relating the quantities, weights and values of purchases.

To encourage the participation of all students, the class was divided into two groups, chosen by the teacher and the Libras interpreter. This division was made by trying to balance the number of deaf students and hearing students. Each group had a sign language interpreter to provide support during interactions. In this way, it was ensured that students had the necessary conditions to communicate with each other.

As part of a different project, the art teacher sought to relate art and mathematics. The task aimed at discussing with the children the characteristics that define a polygon. Pieces of cardboard with different shapes were given to the students so that they could group the pieces using the criteria they considered relevant. The investigation was based on the search for these criteria.

In this task, the students also worked in groups, but this time, formed by free choice. In total, five groups were formed, three composed only of hearing students and two composed of deaf and hearing students. These two groups had the presence of the sign language interpreter. The task provided the students with an opportunity to discuss not only the properties of polygons, but also other topics related to geometry and sign language.

Both activities were unpredictable. In the first activity, we did not know in advance the fruits that students would choose nor the price they would pay for them, making the answers to the questions asked unknown. In the second activity, we did not know the criteria that the students would choose or what discussions could arise from this proposal. We were faced with investigative tasks.

The organisation of the class into groups, as well as the presence of sign language interpreters, provided support for dialogic interactions to take place, in order to contribute to the construction of equity. In what follows, we discuss some excerpts from these classes in which it was possible to perceive the dialogic interaction between the students.

3. The Dialogue among Deaf and Hearing Students

From the tasks performed, we built classroom episodes[1] that provide descriptions of the students' engagement in carrying out the proposed tasks, as well as highlighting the interactions among them. In what follows, we share excerpts from these episodes that support reflections that appear later in this chapter.

In these excerpts, we represent the facial and body expressions that make up part of a sign with sentences in an exclamatory or interrogative format; explanatory comments are represented in parentheses and the translation from sign language to Portuguese is written in square brackets.

We developed tasks with different possibilities of approaches to be followed, which encouraged the sharing of perspectives through dialogue. All the planning was done with the consideration that there always would be a sign language interpreter in the room in mind, even though most students could communicate in Libras. Ensuring this presence was a way to encourage the engagement of everyone in the dialogue. The following excerpt is an example of the sign language interpreter's participation.[2] All names are pseudonyms.

---

1   Full episodes can be seen in Moura (2020).
2   More details on the role of the sign language interpreter in a landscape of investigation can be found in Moura and Penteado (2019).

Giovana: Would you add any more fruit to the salad? If so, which one? (Reading one of the questions.)
Bruna: I think papaya.
Giovana: We were going to buy papaya, but it was too expensive.
Interpreter: [In the fruit salad, was there any fruit missing? Which one?] (Talking to Fábio and Mariana.)
Fabio: [Pear.]
Giovana: Papaya!
Interpreter: [Which fruit?]
Fabio: [Watermelon.]
Mariana: [Kiwi... It was awfully expensive.]
Fabio: [Pineapple.]
Bruna: Pineapple! Pineapple was missing. (Talking to the interpreter.)
Mariana: [Kiwi fruit.] (While trying to get Bruna's attention.) [K-I-W-I take note there.]
Interpreter: Kiwi, Mariana is speaking.
Bruna: So, write there Giovana, papaya, pineapple and kiwi.

The following excerpt shows the collaboration among students in deciding on a presentation to the whole class. In this excerpt, students are interacting collaboratively.

Interpreter: [Can you explain in Libras?]
Valentim: [I speak?]
Interpreter: [Yes, explain to the group.] People, look at Valentim.
Valentim: [Several lines, triangles, and curves.]
Bruna: Let me see if I understand, he said that... he is going... (Speaking to the interpreter.)
Interpreter: Look at him Bruna, he is going to say it again (Interrupting Bruna's speech).
Valentim: [Straights, triangles.]
Bruna: [No! You need to say that you are Group One.]
Valentim: [One!? Why Group One?] (Questioning the interpreter.)
Interpreter: [Because you split into two groups; a group with straight shapes and another with curved ones.]

Barbara then called Fábio and proposed that he explained to the group about straight lines.

Barbara: [You explain the example and I will speak.]
Fabio: [How?]
Barbara: [As Valentim said, Group One, straight lines, triangle... get it?]

After Barbara explained once more, Fábio said he understood what his colleague was saying.

The hearing students, Bárbara and Bruna, are working with deaf students Fábio and Valentim on how the presentation could be performed. The encouragement and assistance of the hearing colleagues was essential for the deaf students to be able to express their perspectives. Such action, to present in Libras, shows concern for the other deaf students in the classroom. In addition, it is an action that helps to naturalise the linguistic differences that exist in the classroom, strengthening dialogic communication between everyone.

Dialogue can also be identified in classroom episodes, through more specific characteristics, such as the different investigation processes. In the following excerpt, the students were asked to find the percentage of each fruit in the fruit salad they had prepared.

> *Teacher:* So, you have to find out what is equivalent to 100%. Do you have the quantity of each fruit?
> *Bruna:* Yes!
> *Teacher:* So, how can we find out the entire amount of the salad?
> *Bruna:* Calculating all these here?
> *Teacher:* Yes, it is a calculation... But which calculation?
> *Giovana:* Division?
> *Bruna:* Multiplication?
> *Teacher:* We want to know the total number of cups...
> *Mariana:* [Sum.]
> *Teacher:* Mariana is right, we will have to add all the cups.

The teacher asked the group to do the calculation and asked them to think of a strategy on how to find the percentage. She made this suggestion and walked away from the group, and Gustavo finished making the calculations. The interpreter asked him to help deaf colleagues who were in some doubt about the quantities that contained decimal numbers.

> *Interpreter:* [You pay attention and Gustavo will help you.] (Talking to Mariana and Fábio.)

Through this interaction with Gustavo, they were able to find the total amount of used fruit cups. Then the students began to think of a way to find the percentage of each fruit. They asked the teacher for help.

> *Teacher:* You guys already found the total number of cups—that was 14, right? Now you need to find a numerical relationship between the amount of each fruit and the total amount. You need to do this to be able to reach the percentage.

Fabio: [Oh I know, 14 equals 100%.]
Interpreter: [That is right.] (Talking to Fábio) Did you understand? (Talking to the other students.)

In this excerpt we observe deaf and hearing students, interpreter and teacher in an investigative situation. The teacher's speech was translated for the deaf students by the sign language interpreter. This helped deaf students to understand what had been said, but it was also a stimulus for collaboration among group members—for example, when she asked the hearing student Gustavo to help his deaf colleagues Fábio and Mariana in carrying out some calculations.

It is noteworthy that a landscape of investigation involves matters that go beyond the content of mathematics. The following excerpt provides an example in which students investigate the meaning of a word unknown to them.

Bruna: Question 9... Does eating fruit salad bring health benefits? ... Benefit? What is this? (Questioning the interpreter.)
Gustavo: Benefit is something that is good for you...
Interpreter: What is benefit? How about searching the dictionary? (Interrupting Gustavo's speech.)

While Bruna searched the Portuguese language dictionary, the interpreter translated the question for the deaf students. But she did not know how to say that word in Libras; she spelt out the word letter by letter in finger signs and asked deaf students to look up the benefit sign in the Libras dictionary.

Interpreter: [We need to know the sign for B-E-N-E-F-I-T. look for it in the dictionary.]
Fabio: [Which word?]
Interpreter: Adriano, please spell the word for them.

And while the deaf colleagues searched for the sign, Bruna read the meaning of the word in the dictionary for the hearing colleagues. When she finished reading, the interpreter questioned her.

Interpreter: Summarising, what does all of this mean?
Gustavo: That is good for the skin and the body?
Giovana: No!
Bruna: Look, it is this one, benefit! (Pointing to the word in the dictionary.)

> *Giovana:* Let me see, Bruna (Taking the dictionary from Bruna's hands) ...Strange!
> *Gustavo:* Let me try to understand... The word is simple. (Holding out his hands asking Bruna for the dictionary.)
> *Interpreter:* Read aloud to see if the girls can understand.

After this reading, the interpreter discussed with the students the meaning of this term, which was interpreted by them as things that are good, that do well. Even though they understood the meaning, it was not possible to find a sign in Libras for that term during the task. The interpreter promised to research and bring the sign relating to the word *benefit* in the next class.

In a mathematics lesson, the words used to refer to mathematical terms or concepts are not always known, nor do they always have a defined sign in Libras, so there must be space in the class to explore this. In the excerpts above, students seek to understand the meaning of the word *benefit*. Such investigation was not foreseen in the task planning, but it was important for the students to be able to proceed with the task. In addition, it expanded the students' vocabulary.

This expansion of vocabulary is extremely important when considering a learning environment with the presence of deaf students. As Libras is a developing language, there are many signs that are built from the demands and production of new knowledge by students. Thus, it is common that in mathematics lessons there are negotiations of the signs for concepts newly presented to students. In the task involving the concept of the polygon, defining a sign for such a concept proved to be necessary, so the interpreter conducted the search for a sign referring to the polygon.

The interpreter invited Valentim to go to the blackboard and explain what he was thinking. The student showed some uncertainty when trying to define a sign for polygon, and Pablo volunteered to go to the blackboard to help him. Pablo pointed to a triangle, and then made the classifier referring to the triangle but Valentim interrupted him:

> *Valentim:* [No! Is the sign.]
> *Interpreter:* [We want a sign for this word.] (Pointing to the word polygon, written on the blackboard.)

The hearing students were also interested in defining the sign, and Bruna alerted the interpreter that she had an idea. The interpreter, Ana, asked her to wait and continued the conversation with Valentim and Fabio:

> *Interpreter:* [Polygon means many sides, many lines.]
> *Valentim:* [Many straights.]
> *Pablo:* [Many stars!]
> *Interpreter:* [No star, it is straight!]

Valentim then suggested making the sign for "many", followed by the line sign, to represent the straight lines. The interpreter thought that would be a good sign but decided to listen to Bruna as well.

> *Interpreter:* Okay Bruna, you can give your opinion.
> *Bruna:* Ahhh... Do the p sign and the square sign.
> *Teacher:* But a rhombus is also a polygon, so what?
> *Bruna:* Ahhh... Never mind!

At this time, the interpreter asked the deaf students to collaborate with Pablo and Valentim. From the sign suggested by Valentim and the idea given by Bruna, Pablo created a new sign. First, he made the sign for "many" and then, with the hand configuration in P, the sign referring to the lines.

This negotiation was unpredictable; however, it was essential for students to understand the concept of polygon, not just the spelling. Negotiations of this nature are important for the expansion of Libras in the lexical field. Furthermore, it is an opportunity to encourage collaboration among students in exploring the properties of mathematical objects.

By positioning themselves at the front of the class and sharing their perspectives, deaf students became the protagonists of the investigative process, and consequently of their learning processes. Following this excerpt, all students were invited to collaborate in the construction of a sign for polygon. Deaf and hearing students gave their opinions, aiming to choose the best sign, thus establishing a relationship of equity, since all students could collaborate on an equal footing. In this sense, we highlight the potential of dialogue in building an equitable relationship between the participants.

The group started looking at the fruit stands. Adriano, Fabio, and Gustavo stopped near the watermelons and were amazed at the price and called the rest of the group over.

> *Adriano:* Look, it costs R$1.89.

Giovana: Wow! A great price.
Adriano: Oh... They inverted, Giovana... It is the whole fruit by R$1.59 and a piece is R$1.89...
Giovana: Hey... A piece R$1.89?
Adriano: Yes, that is what is written here...
Fabio: [Watermelon is good, and the price is good.]
Interpreter: What is missing for buying?
Bruna: Look, we have already bought bananas, oranges, strawberries, apples and... What else is there on the Pietra list?
Pietra: Let me see?
Bruna: Wait...Now we are seeing the price of the watermelon and seeing what fruit we can take.
Interpreter: Watermelon?
Bruna: Yes! The price is good.
Mariana: [I think watermelon does not match.]
Bruna: Will we choose that? It is better to look for another fruit (agreeing with Mariana).

The group followed Bruna and Mariana through the fruit stands until everyone gathered around the stand where the grapes were.

Bruna: What do you think about this one? (Pointing to a type of grape.)
Fabio: [This one] (Showing a large bunch of grapes of another type.)
Mariana: [No! This one has much, and it is expensive, better this one.] (Showing the grapes that Bruna had referred to.)
Giovana: Ah, but this one (referring to the bunch of grapes shown by Fabio) I think it goes better with the fruit salad.
Bruna: It is true! Let us take this. (Referring to the bunch of grapes suggested by Fabio.)

In this excerpt, we see the interaction of a group buying fruit at the supermarket. Students were finishing shopping and needed to make decisions on which fruit they could still buy with the rest of the money. The arguments of the deaf students Fábio and Mariana have the same validity as the hearing students—that is, they actively participated in the decision-making carried out by the group. That hearing students knew Libras was especially important. This favoured the construction of equity among all students. However, we do not think this reduces the importance of the role of the sign language interpreter.

From the highlighted excerpts, we believe the tasks favoured the inclusion of deaf students in the learning processes, a cooperation that allowed all students in the class to have equal conditions to participate in the proposed investigations. The support from the sign language interpreter allowed collaboration among the students, as well as making room for new investigations during the task. Different learning outcomes emerged through these interactions—for instance, the meaning of word *benefit* and a sign for polygon.

Language was the main difference among the students. The appreciation of Libras (not the dominant language) in this context was crucial to the construction of equity among students since it facilitated the engagement of all students in a dialogue. Thereby, we claim that these tasks substantiate the proposal of inclusive landscapes of investigation (Skovsmose, 2019).

4. Final Remarks

In this chapter, we shared some reflections about mathematics classroom episodes in which deaf and hearing students were brought together. We highlight the organisation of investigations based on dialogic communication, which was essential to face unpredictability and enable collaboration in conditions of equity in the classroom.

For Moura (2020), dialogic communication supports the encounter of differences. To think of inclusive education in terms of meetings is to think of an education that recognises and values each one, and that envisions the possibility of learning from what is different.

In these investigations dialogic communication facilitated the meetings amongst deaf and hearing in the classroom. Other meetings also took place—for example, between sign language interpreters and hearing students. These meetings do not only refer to sharing the same space, but to a movement of seeing the other, of wanting to be together, of favouring cooperation and the construction of equity, essential elements for the inclusion of deaf students in mathematics classes.

Finally, we understand that the reflections presented in this chapter support the proposal of inclusive landscapes of investigation, bringing an important perspective to the teaching and learning mathematics to students with disabilities. Encouraging collaboration, collective thinking and dialogic communication in order to learn something new is a way to support a mathematics education practice addressed to differences. We hope that this chapter can inspire teachers in their practices towards a mathematics education for all.

References

Alrø, H. & Skovsmose, O. (2004). *Dialogue and learning in mathematics education: Intention, reflection, critique.* Kluwer Academic Publishers.

Figueiras, L., Healy, L. & Skovsmose, O. (2016). Difference, inclusion and mathematics education: Launching a research agenda. *International Journal for Studies in Mathematics Education, 9*(3), 15–35.

Milani, R. (2015). *O processo de aprender a dialogar por futuros professores de matemática com seus alunos no estágio supervisionado.* Doctoral dissertation. Universidade Estadual Paulista (Unesp).

Moura, A. Q. & Penteado, M. G. (2019). The role of the interpreter of Brazilian sign language in the dialogue among deaf and hearing students in mathematics classes. In D. Kollosche, R. Marcone, M. Knigge, M. G. Penteado & O. Skovsmose (Eds), *Inclusive mathematics education: State-of-the-art research from Brazil and Germany* (pp. 253–270). Springer.

Moura, A. Q. (2020). *O encontro entre surdos e ouvintes em cenários para investigação: das incertezas às possibilidades nas aulas de matemática.* Doctoral dissertation. Universidade Estadual Paulista (Unesp).

Skovsmose, O. (2001). Landscapes of investigation. *ZDM—Mathematics Education, 33*(4), 123–132. Reprinted as Chapter 1 in O. Skovsmose (2014), *Critique as uncertainty* (pp. 3–20). Information Age Publishing.

Skovsmose, O. (2019). Inclusions, Meetings and Landscapes. In D. Kollosche, R. Marcone, M. Knigge, M. G. Penteado & O. Skovsmose (Eds), *Inclusive mathematics education: State-of-the-art research from Brazil and Germany* (pp. 71–84). Springer.

13. Inclusion and Landscape of Investigation:
A Case of Elementary Education

Íria Bonfim Gaviolli and
Miriam Godoy Penteado

> This chapter presents aspects related to inclusion, landscapes of investigation and the perspective of *deficiencialism*. These ideas are based on a Master's study of data produced in an elementary school classroom in the city of Rio Claro, in the state of São Paulo (Brazil). In this classroom, mathematical tasks from the perspective of a landscape of investigation were developed. We present a brief contextualisation of the research, methodological aspects and theoretical assumptions on inclusive education, as well as some of the children's dialogues during the development of the tasks. As a conclusion, we present considerations on the relevance of the landscapes of investigation approach when considering inclusion from the perspective of *deficiencialism*.

In this chapter we discuss aspects of inclusion, landscapes of investigation and the perspective of *deficiencialism*. The discussion is based on the results of a research study whose objective was to search for elements that could favour the engagement of a student with Autism Spectrum Disorder (ASD) in mathematics classes organised around landscapes of investigation (Gaviolli, 2019).

The data production took place in a classroom of children between seven and nine years old across eight lessons in which a landscape of investigation was proposed. This classroom included one particular student, Dani (pseudonym), diagnosed with ASD. The tasks were carried out by one of the researchers (the first author of this chapter), assisted by the classroom teacher. At the beginning of the data production, the researchers' attention was focussed exclusively on this student, Dani—however, during the lessons, relevant issues caused by the presence of other students in the classroom arose.

For data analysis we used the notion of *deficiencialism* developed by Renato Marcone (2015, 2018). Marcone was inspired by the idea of orientalism to create the term—that is, inspired by the conception of the east by the west—in order to problematise the invention of disability by someone who believes themself to be within a pattern (also invented) considered as normal. As a consequence of using this terminology, a dichotomy emerges between "normal" people and people with disabilities. This dichotomy is marked by a distinction that refers specifically to a corporal difference as "the external, superficial difference, the first to be noticed when there is contact with the other" (Marcone, 2015, p. 18). The dichotomy is based, therefore, on the invention of an atypical condition which, due to a stipulated standard of "normality", becomes classified as being inferior. From the perspective of deficiencialism, one can even problematise the invention of certain identities.

Therefore, we emphasise that, in our description of classroom interactions, we are not setting out Dani's identity according to a normalising discourse, regulated by means of medical diagnostic criteria and characteristics provided in a report or in the Diagnostic Manual of Mental Disorders, DSM-5 (APA, 2014). We interacted with Dani (as we came to know her) during a period of familiarisation—a time when we were at school to meet and join in the routines of the teacher and students—and from conversations with the teacher responsible for the specialised educational service in the school. Thus, we emphasise that, from now on, when talking about elements of Dani's engagement, we are thinking about her as an integral part of a classroom, and not as an invented identity. Dani was who she was, being with João, Miguel, Rayssa and Antonio (all pseudonyms). That is why it is important to consider all students in the classroom in order to fully examine their interaction during the proposed tasks.

1. Landscapes of Investigation and Mathematical Tasks

Ole Skovsmose (2000) emphasises that the idea of a landscape of investigation presupposes that students participate in an investigative process by asking questions, planning lines of investigation, engaging in an exploration process and using justified argumentation. In a landscape of investigation, students are able to ask questions that would otherwise be asked only by the teacher. Questions such as: "What happens if ...?", "Why is it that way ...?" are no longer just the teacher's attempt to engage students. They characterise the students' interest and involvement in the tasks (Alrø and Skovsmose, 2004; Skovsmose, 2001, 2011). It was these ideas that guided the proposed tasks for the students who participated in our research.

For us, thinking about the organisation of mathematical tasks in this way allows students to be introduced to a classroom organisation different from that based on a style of teaching in which the textbook and the teacher represent the highest authorities. Adopting this approach allows us to take on a pedagogy different from a conventional pattern, where some content is presented, followed by examples, and the requirement of each student is merely to solve exercises, which most of the time have a single correct answer. This type of organisation in the classroom is located in what Skovsmose refers to as an "exercise paradigm".

In such an investigative process, some patterns of communication that are commonly present in the classroom can give way to others. Alrø and Skovsmose (2004) point out the sandwich pattern of communication as one such pattern that can be different in a landscape of investigation. The sandwich pattern is characterised by a communication model in which the teacher asks, the student answers and the teacher validates the answer. Bureaucratic absolutism is another feature accompanying the exercise paradigm. It is characterised, above all, by the same treatment of different types of errors in relation to mathematical exercises. Errors in interpretation of a question, algorithms and mathematical concepts are all treated as absolute, without the teacher pondering any explanation or argument by the students.

Skovsmose emphasises that a change in the learning environment—from exercise paradigm to landscape of investigation—does not mean that exercises need to be eliminated from mathematics classes, but instead that the way the tasks are conducted need to be changed. For example, instead of presenting a classification of polyhedrons to the students based on their number of faces, edges and vertices, students can be invited to investigate what they might have in common, and only after this exploration would the categorisation be presented. We are interested in hearing what students have to say: What will their criteria for categorisation be and how will they organise the common elements? We will be interested in analysing the differences between the students' answers.

Below, we present some tasks performed in the classroom, showing some strategies that contributed to the students' engagement. We highlight some situations and reflect on them inspired by the concept of *deficiencialism*.

Situation I: One of the tasks performed by the students was to identify the names of geometric figures. For this, we presented plane shapes in the form of circles, triangles, rectangles and squares.

Dani was still learning to read and write so she was only able to recognise some syllables, but had difficulty using them to make words. We presented the shapes saying in a loud voice (for example): "This is the square." It also happened that some students had already recognised and uttered the name of the shape. In situations like this, it was common for everyone to say the name of that shape at the same time.

> *Rayssa*: Ahh! This is the circle.
> *Giovana*: Yes, it is the circle.
> *Some voices in chorus*: Circle, circle, it is a circle.
> *Dani*: Circle.
> (Gaviolli, 2019, p. 70, our translation)

We consider that two of the strategies that helped Dani's engagement in this task were: (1) the use of visual resources, and (2) presenting the written name of plane figures, because she was also in the process of learning how to read and write. She was able to associate the sounds of some of the written names with the visuals. For this, once we had completed the presentation of names and shapes, we came to Dani so that she could manipulate the material. We presented two or three options of names from which she could choose one that she thought was the name of the piece on her table. Then, we said the name of the shape and she recognised the syllables, joining the printed name with the representation of the plane figure. Other times, we would show a piece and ask her to choose, from printed words, which could be the name of that piece. Figure 1 is an illustration of her table.

Fig. 1. Dani's table. Source: researchers' personal collection.

Situation II: Children mentioned the names of everyday things they liked and we tried to relate these to some geometric solids. For example, Arthur and José liked to play the videogame Minecraft, so we related the geometric representation of the cube with the Minecraft design. During lessons, whenever we referred to the term 'pyramid', John remembered the pyramids of Ancient Egypt. Dani, in turn, showed an interest in dinosaurs, so we looked for images that contained dinosaurs and geometric solids. When we started the activity showing a slide with initial pictures of an ice cream cone and old pyramids, the children were excited and said: "Ahhhh!", "Ohhh!", "Look!".

Dani: Look! The dinosaur!
José: Look! Minecraft!
Rayssa: Look! The ice cream cone.
José: The pyramid that is in the desert.
Dani: Look! The dinosaur! [At this point, she got up and went to the front of the room to point at the dinosaur drawing on the slide.]
José: Look! Minecraft! [At this point he got up and went to the front of the room to point at the Minecraft drawing.]
[...]
Rayssa: The one in front of the dinosaur looks like the big cube of golden material [a material they use for calculation].

Dani: I saw the ice cream cone.
Giovana: Íria, the ice cream cone looks like a unicorn horn.
[...]
Rayssa: The one over there, the magic cube, looks like the golden material cube.
Arthur: The square behind the dinosaur looks like a die, but it has no colours.
Íria: Is it a square that is behind the dinosaur?
Arthur: It seems to be.
Dani: It is a square.
Íria: Shall we remember its name? We just saw it. What category is it in?
Some children: Cube.
[...]
[On a slide, a photo of the Louvre Museum appears.]
Nicolas: Look! That pyramid is in Paris, France.
Arthur: In the video game there are countries above. Below there is Paris and these football teams that appear. That is why he knows.
Íria: How cool, boys!
João: There are several types of pyramids.
Íria: What do you mean, João?
João: Some pyramids are made of glass, others are made of bricks.
Íria: I understand, João... Guys, this pyramid in Paris is a museum. Do you know what a museum is?
Giovana: A museum is where you keep old things. I have never been to one.
[At this point they all start talking at once.]
José: I have been to a museum.
Giovana: My aunt has already been to that museum there.
(Gaviolli, 2019, pp. 70–71, our translation).

We understand that the simple fact of showing something that was of interest to students stimulated them to talk and interact with each other. It was a way for them to be engaged with the activity.

Situation III: Here, the idea was to ask the students to draw objects they found in the school. The objects should remind them of some of the geometric plane shapes. For this task, we made the models of the plane shapes available on the tables. During the task, we realised that they took those pieces in their hands, looked at the sides, looked up and down and said: "I don't know what to draw." So, we suggested that they thought of the school environment. It could be anything. After a while, they started drawing. In this task, the specialised educational service[1] teacher was in the classroom and put some questions to Dani: "Dani, which shape reminds us the cover of this book?" (pointing to the shapes on the table), "And that clock on that wall, which one is it similar to here?" From this conversation Dani became inspired to create her drawings.

Situation IV: After having introduced the students to the names of geometric solids in the previous tasks, we invited them to build their own solids using Play-Doh, which is a modelling compound.

We realised that we could use this resource as an ally for the development of the proposed task, given the fact that the whole class liked it and Dani, in particular, had very good modelling skills. The specialised educational service teacher once told us that while Dani was still in preschool, students lined up around her desk asking her to make models of dinosaurs. During the classes that we used Play-Doh in, we noticed that at times children also went to Dani when they wanted to model something different, like a dinosaur, for example.

Our approach was to use something of common interest to the students. In this case, it was Play-Doh, but it could have been any other resource.

Situation V: In one of the lessons, we told a story—something that their teacher regularly did, and that they enjoyed. We used images during the narration in order to have a visual representation of what was happening.

One of the stories was about the Ugly Cylinder (Cilindro Feio), which we illustrated with photographs. In this meeting, Dani was not feeling well, and at first she did not seem to want to accept the invitation to be engaged in any conversation. However, while we were telling the story, she started to interact with the images that appeared on the slides.

> *Íria*: Had he [the Ugly Cylinder] found his partner?
> *José*: No.
> *Most children*: Yes.
> *Dani*: Look! Look, she is yellow. [At this point we had passed a slide, and Dani exclaimed about the colour of Mrs Cylindra's [Dona Cilindra's] clothes.]

---

1   This term describes the use of Special Education resources to support inclusion in the regular school system.

> [We continued telling the story. At one point the Ugly Cylinder met with Mrs Pyramid (Dona Pirâmide).]
> *Dani*: He liked it.
> *Íria*: He liked her.
> (Gaviolli, 2019, p. 74, our translation).

Telling a story allowed us to talk about other subjects. For example, in a certain part of the story when the Ugly Cylinder met Mrs. Pyramid, she asked why he was so sad. He, crestfallen, replied that he was clumsy, had no edges, no vertices and rounded bases and therefore he could not find a match. At this point in the story, the researcher intervened, making some comments, and Rayssa and Giovana participated by exposing their points of view:

> *Íria*: It is important to respect colleagues, their differences. Each one has his or her own way of being.
> *Rayssa*: It is true. Everyone has their own way, because some people are blond and some have black hair, blue hair.
> *Íria*: That's right, Rayssa.
> *Giovana*: Íria, other people are also disabled.
> *Rayssa*: Some have short hair, some don't.
> (Gaviolli, 2019, p. 74, our translation).

Situation VI: Here we indicate a general aspect that we noticed throughout the lessons. During the first classes we were always insisting on Dani's participation; sometimes she interacted and sometimes she did not. We asked for some advice from the specialised educational service teacher, who had known Dani since she was small. She advised us that whenever we asked a question, we should offer some options for answers. Thus, when asking for the name of an object, we then offered two or more options from which Dani could choose an answer. It is worth saying that we also used these strategies for other students who remained silent during lessons. Doing this, we realised that when we offered these options, the dialogue flowed better—that is, in most cases there was some interaction and, from that interaction, we could continue the conversation. This was essential for the students' engagement in the process. For us, it was important to listen to everyone in order to understand what they were thinking and conjecturing.

2. Reflections on Landscapes of Investigation and Deficiencialism

In this section we present some reflections on the landscape of investigation and the perspective of *deficiencialism*.

A first consideration is about the different tasks and strategies that made it possible for students to move through different learning environments. The task and strategies concerned speech, writing and deciding which materials were to be made available. We recognised that the different tasks and strategies made students feel invited to participate in the proposed activities.

In order to clarify what we mean, we will return to the episodes from the classroom. We rarely heard the voices of some of the children; however, this did not imply that they did not accept the invitation to engage, because when asked directly, they responded. In addition, they were always willing to participate in tasks that involved some written production. On the other hand, there were those children who liked to participate in the conversation, yet showed some resistance when it came to doing written tasks. We realised that, by making different strategies available, students could move through the different learning environments and participate in the proposed tasks according to their condition.

The movement between different learning environments also made it possible for us to expand the resources available for carrying out the mathematical tasks. During classes, for example, we used: Play-Doh, letters printed on paper, slides and geometric solids of different colours, shapes and sizes. There was a diversified repertoire that supported the development of tasks.

As a result, we highlight the importance of moving through different learning environments. This makes it possible to diversify types of task and strategy, in order to favour pupil differences present in the classroom. We also emphasise that both the tasks proposed in the paradigm of exercises and those in the landscape of investigation were important for the students' learning. When going through the different references of the learning environments, they had the possibility to experience different tasks and patterns of communication.

A second consideration, which can be seen as a consequence of the first one, concerns the possibilities of conversations that took place during the performance of tasks. By offering some strategies for carrying out the tasks, we were able to contribute so that the students could expose what they were thinking and, from that, establish communication with other students and with the researcher. For us, it was important to offer ways for all students to be able to explain what they were thinking about the topic.

We understand that prioritising a range of possibilities for tasks and strategies—which is what results from moving to different learning environments—converges with the perspective of Marcone (2015). If means are offered so that certain tasks can be performed, the person is no longer *disabled* in relation to a given context. In other words, we establish an approximation of a landscape of investigation with *deficiencialism* because, by offering different conditions for students to participate in the tasks, the limitations of the students classified as disabled might disappear. It is important to emphasise that it is not our intention to create a method or even to say that the situations mentioned by us would be directly applicable to other contexts. Rather, we want to share how we find it possible to include all the students in what we do in the classroom. Considering the *deficiencialism* perspective, students have the possibility of leaving the condition of being disabled. It is also important to mention that it is not a case of a student leaving the condition of disabled person behind in order to become a *normal* person. It is about leaving behind the situation of the classroom in which the students are seen, mainly, by what is lacking in them.

Therefore, when thinking about the landscape of investigation combined with the perspective of *deficiencialism*, we must consider expanding the repertoire of tasks, conditions and learning environments, as well as encouraging movement through these environments. With that, we understand that different communication patterns can emerge and that opportunities can be offered so that all the students in a classroom have the possibility to engage with the proposed tasks.

Acknowledgements

This text is a short version of one of the chapters of Íria Gaviolli's Master's thesis (2019) entitled: Cenários para investigação e Educação Matemática em uma perspectiva do deficiencialismo [Landscapes of investigation and mathematics education from the perspective of deficiencialism].

References

Alrø, H. & Skovsmose, O. (2004). *Dialogue and learning in mathematics education: Intention, reflection, critique*. Kluwer Academic Publishers.

American Psychiatric Association (APA) (2014). *Manual diagnóstico e estatístico de transtornos mentais: DSM-5*. Artmed.

Gaviolli, I. B. (2019). *Cenários para investigação e Educação Matemática em uma perspectiva do deficiencialismo*. Master's thesis. Universidade Estadual Paulista (Unesp).

Marcone, R. (2015). *Deficiencialismo: a invenção da deficiência pela normalidade*. Doctoral dissertation. Universidade Estadual Paulista (Unesp).

Marcone, R. (2018). Descontruindo narrativas normalizadoras. In F. Rosa & I. Baraldi (Eds), *Educação matemática inclusiva: estudos e percepções* (Vol. 1, pp. 17–36.). Mercado das Letras.

Skovsmose, O. (2001). Landscapes of investigation. *ZDM—Mathematics Education, 33*(4), 123–132. Reprinted as Chapter 1 in O. Skovsmose (2014), *Critique as uncertainty* (pp. 3–20). Information Age Publishing.

Skovsmose, O. (2011). *An invitation to critical mathematics education*. Sense Publishers.

14. Landscapes of Investigation with Seniors[1]

Guilherme Henrique Gomes da Silva,
Rejane Siqueira Julio and
Rafaela Nascimento da Silva

The authors discuss how seniors engage in mathematical activities when inserted into landscapes of investigation. Through a qualitative approach, the researchers collected data in two meetings with a group of seniors who were involved in landscapes of investigation in a mathematics education project aimed at that age range. The authors also highlight that the activities fostered interaction/collaboration between the seniors, causing them to change their conception of mathematics, producing knowledge and raising philosophical, aesthetic, technological and mathematical discussions.

Several countries around the world are experiencing the phenomenon of an increasingly ageing population. In Brazil, according to data from the Brazilian Institute of Geography and Statistics more than a third of the population will be aged 65 and over in three decades' time (IBGE, 2019). This phenomenon has been one of the main reasons for governments to act in favour of ameliorating the quality of life of this growing section of the population—for example, through enhancing educational opportunities.

---

1   This chapter is adapted from Silva et al. (2019).

In the case of Brazil, such practices are expressed as a right of seniors in the Legal Rights of the Senior Citizens (Law n. 10.741 of October 2003) and in the National Education Plan (Law n. 13.005 of June 2014). Additionally, research in education—and in mathematics education in particular—has shown that the involvement of seniors in such practices can help them remain cognitively active, interact socially and learn new content. It can also help them establish the necessary knowledge to claim their rights and participate more actively, critically and creatively in their own lives, remaining active in the community (Julio and Silva, 2019; Lima, 2015; Lima et al., 2019; Scortegagna, 2010). In addition, Scagion (2018) warns that the vision of seniors related to mathematics may be connected mainly to common sense conceptions and ideas. Based on Social Representation Theory from Moscovici (2005), Scagion (2018) has interviewed seniors who have participated in mathematics activities offered by a university project. "Mathematics is everywhere" or "Mathematics is important for quality of life" were some of the social representations identified among the participants. Scagion (2018) points out that these representations end up being incorporated in their discourse in only a very superficial way. They are not able to provide details about the nature of usefulness and the contexts where mathematics could be applied. According to Scagion, involvement in educational practices related to mathematics education could expand their repertoire.

In this chapter, landscapes of investigation with seniors[2] will be analysed, bringing together elements from our experiences in two meetings held in 2018 through an extension project[3] aimed at the development of mathematics education for seniors, called Mathematics Conversations. These meetings were organised around two topics: the golden ratio and the Fibonacci sequence. In the first meeting, eight participated and in the last, seven participated. Table 1 introduces the seniors that participated in both activities. All names are pseudonyms.

---

2   In Brazil, a person considered to be "senior" is over sixty years old.
3   Extension Projects are understood as educational, cultural, and scientific processes that articulate the scientific knowledge with the community outside the university.

Table 1. Seniors participating in activities.

Name	Age (years)	Education	Meeting 1	Meeting 2
Lúcia	67	Portuguese Teacher	X	X
Marina	65	Pharmacist	X	
Simone	63	High school education	X	X
Sandra	65	Psychologist	X	X
Joana	63	Pharmacist		X
Selma	64	Physical Education teacher		X
Lúcia	67	Attorney	X	
Glória	71	Teacher training degree	X	X
Ana	65	Mathematics teacher	X	
Pedro	65	Electrician	X	
Adelia	65	Geography teacher		X

The research method was participant observation. Data was collected through audio recordings of the meetings and conversations with the seniors during and after the activities, and field notes. The research team transcribed and organised the data to analyse how the seniors engaged in landscapes of investigation. The processes of preparing landscapes of investigation, as well as the analysis, are based on Skovsmose (2000, 2011) and on research in mathematics education with seniors, such as Lima (2015), who provided relevant guidelines for working with this age range.

1. Landscapes of Investigation with Seniors: The Mathematics Conversations Project

Lima (2015) was a pioneer in the mathematics education field involving systematic approaches to the development of pedagogical activities aimed at seniors. His research inspired us to develop the *Mathematics*

Conversations project, aimed at seniors participating in the Third Age Open University Program at the Federal University of Alfenas. Since 2018, activities have been held weekly at this university, and include the active participation of seniors with different backgrounds—as illustrated by Table 1—who chose to participate in the project to learn more about mathematics and stimulate their minds. Two teachers and four students from the mathematics degree programme at the Federal University of Alfenas formed the project team in 2018. The team planned the activities weekly and discussed what occurred in previous meetings with the seniors. In this process, the team explored articles and experiences, and discussed the adaptation of teaching materials for working with seniors. In addition, doubts and mathematical questions raised by the seniors were used as possibilities for the creation of new proposals. The project also provided research data related to mathematics education and seniors.[4]

The project sought to develop activities in a learning environment that would foster the creation of landscapes of investigation, as proposed by Skovsmose (2000, 2011). The landscapes of investigation differ from activities based on the "exercise paradigm", in which the central premise is that for each exercise or task, there is one, and only one, correct answer, and the teacher is at the centre of the educational process. A landscape of investigation favours the use of investigations in teaching and learning processes. It makes the students responsible for exploring and explaining/justifying knowledge production, and depends on their acceptance of the investigation. In a learning environment, the differences between the pedagogical work based on landscapes of investigation and the exercise paradigm can be categorised based on the articulation of these two practices and three references (Table 2).

Table 2. Learning environments. Adapted from Skovsmose (2000).

	Exercise Paradigm	Landscapes of Investigation
References to Mathematics	(1)	(2)
References to Semi-Reality	(3)	(4)
References to the Real World	(5)	(6)

---

4   For example, Silva (2020), and Julio and Silva (2019).

The first reference is related to questions and mathematics activities. The second, a constructed reality, is called semi-reality, and the third refers to situations in the real world. In school mathematics education, Environments 1 and 3 are the most common, as they refer to exercises formulated in the context of mathematics and semi-reality, respectively. In Environment 1, imperative technical activities predominate, such as "Solve the equation ..." or "Calculate the value of the hypotenuse of the given right triangle". In Environment 3, situations are constructed in order to train students in certain mathematical techniques. Even though Environment 5 is based on real-world data, students only use such information to solve a given closed task. Environments 2, 4 and 6 are related to the use of investigative practices based on the three references. According to Skovsmose (2000, 2011), in landscapes of investigation, questions such as "What happened if..." characterise the teacher invitation, while questions such as "Yes, what happened if..." characterise students' acceptation of entering the investigation process.

Considering that most of the school mathematics experienced by the seniors will have been based on the exercise paradigm, which in many cases caused traumas and even an aversion to mathematics, we started with landscapes of investigation.[5] This was revealed to be a good option, because it encouraged different discussions and conversations about mathematics and other subjects, captivating the participants and causing more seniors to join the project, which started with only one senior woman. From there, other seniors began to participate, losing their fear of facing mathematics-related matters.

Another aspect is that in landscapes of investigation, the process is more important than a specific result. In this process, there are many conversations about mathematics and other subjects among the participants. The encouragement and valorisation of these conversations are based on a critical conception of education (Alrø and Skovsmose, 2003; Freire, 2016), prioritising dialogue as fundamental to the production of knowledge. Alrø and Skovsmose (2003) provide us with guidelines regarding the use of dialogue with elements that constitute what they call the inquiry-cooperation model (IC-model). This model is formed by dialogic acts that emerge during the interaction between teachers and learners. In our case, with the project team and the seniors,

---

5   See the discussion in Julio and Silva (2019).

these were namely: to establish contact, perceive, recognise, take a stand, think aloud, reformulate, challenge and evaluate. To develop the pedagogical activities, we also considered Lima's recommendations (2015). According to him, it is recommended that we prepare the environment in advance of receiving the seniors—increase font size, use appropriate voice intonation for everyone to hear well and fully participate in the activity, and finally respect the time the senior needs to carry out the activities.

2. The "Golden Ratio" and "Fibonacci Sequence"

To discuss the engagement of the seniors in two landscapes of investigation, we begin by describing how the development of the learning environments occurred. Afterwards, we bring in elements that will enable us to raise the topic of discussion as planned.

In order to work with the "golden ratio" and the "Fibonacci sequence", the project team encouraged the construction of learning environments based on landscapes of investigation. The matters came from a request from one of the seniors, who wanted to know what "golden ratio" meant and what its relationship was with the Fibonacci sequence.

As the matters had arisen from the interest of only one participant, the team was concerned with building learning environments so that all the seniors would accept the invitation to participate. Thus, we planned two activities, implemented in two different meetings, called "golden ratio landscape of investigation" and "Fibonacci sequence landscape of investigation". To plan the activities, we carried out a bibliographical review on these matters.[6] Thus, the team "opened the exercises" (Skovsmose, 2011)—in other words, they transformed/adapted the activities from the bibliographic review to compose the landscapes of investigation.

The golden ratio landscape of investigation was developed with the seniors as follows. First, we showed them several images: Leonardo da Vinci's Mona Lisa; the Notre Dame cathedral in Paris; a shell; the Apple

---

[6] For the study on the theme and elaboration of the golden ratio scenario, the team used works by Garcia et al. (sn) and Queiroz (2007). For the Fibonacci sequence landscape of investigation, works by Lívio (2008), Belini (2015), and Leopoldino (2016) were used.

company symbol; an ear; a sunflower; a credit card, and the Vitruvian Man by Leonardo da Vinci (Fig. 1). Then, we asked the seniors to say what they thought the images had in common. The intention was to invite them to explore and raise questions.

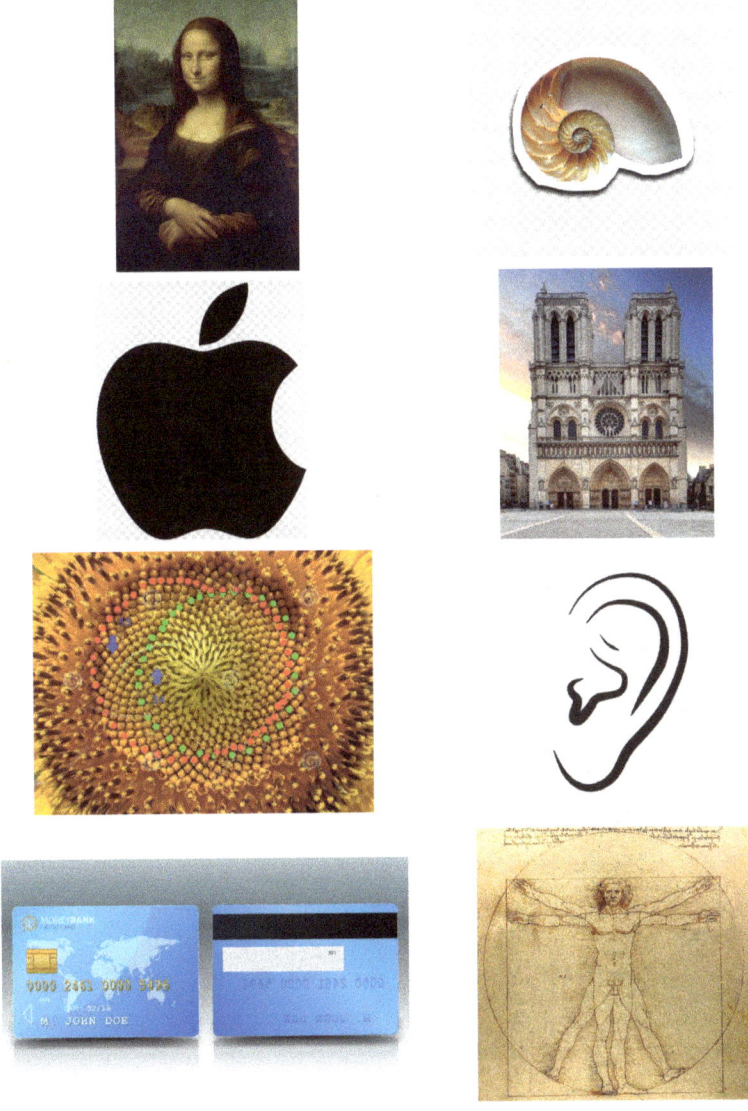

Fig. 1. Images showed to the seniors. Source: the internet.

Following that, we presented the same figures with golden spirals inscribed in rectangles, without telling them at once what that meant (Fig. 2).

Fig. 2. Images showed to the seniors. Source: the internet.

We then proposed a discussion about the golden ratio, addressing its historical context and the mathematical elements involved. We started by inviting a senior woman to read a text,[7] which highlighted the use of the golden ratio in the Renaissance. The text featured a picture of the Vitruvian Man, showing a man of golden proportions inscribed on a circumference and a square. After the questions, comments and hypotheses, we proposed that the seniors should use their own body measurements to explore the golden ratio. In the last step, the seniors constructed the golden rectangle with a ruler and a compass, and then calculated the golden ratio of the rectangle constructed. To conclude, the seniors were given other images, such as swirls, snails, milk glass flowers and the symbols of Toyota, Pepsi and the National Geographic television channel (Fig. 3). In these symbols, the seniors could identify the golden ratio, or else verify that its proportions referred to it.

Fig. 3. Examples of figures presented to the seniors. Source: the internet.

The Fibonacci sequence landscape of investigation sought to make the seniors discuss the functioning of a mathematical sequence. Specifically,

---

7   Based on Queiroz (2007).

it aimed to help them identify a relationship between the mathematical contents handled in the golden ratio scenario with the Fibonacci sequence. As the seniors were eager to know the historical context, the project team chose to introduce the subject by reading a text[8] about Leonardo de Pisa (or Fibonacci), highlighting his importance in the introduction of the numbering system we use currently—the Indo-Arabic numbering system or decimal numbering—and the rabbit problem, which allows one to establish the Fibonacci sequence. The rabbit problem, adapted from Lívio (2008), is formulated as: "A man sets a couple of rabbits in a fenced place. How many couples of rabbits can be generated by this couple in a year if we assume that each month each couple generates a new couple that will start to breed from the second month of life?" After presenting the rabbit problem, the seniors were invited to work on it collaboratively, using a cardboard table and cardboard pictures of adult and young rabbits they could manipulate. The seniors were first required to familiarise themselves with the proposal, which they did by exploring and asking questions, and by establishing negotiations and strategies to solve the problem.

To promote discussion about the results obtained, the seniors completed a table individually by registering the number of couples of adult rabbits, young rabbits and the total number of rabbits for each month. The objective was to establish the relationship between the elements of the sequences formed in the columns of the table to determine the characteristics of the Fibonacci sequence. Then, to relate the Fibonacci sequence to the golden ratio, the team invited the seniors to divide the[9] terms of this sequence and register the results in their table, to investigate what was happening with those ratios.

The seniors were given images to observe, such as the family tree of a drone (male bee), the configuration of sunflower seeds, the number of petals of a daisy flower and the growth of plant branches—all of which can serve to illustrate the Fibonacci sequence. Afterward, the team proposed as homework the resolution of some questions that had been taken from textbooks that dealt with numerical sequences from words and shapes (Fig. 4).

---

8   The elaboration of the text was based on Lívio (2008).
9   The division between the terms was carried out as follows: $F_n F_{n-1}$.

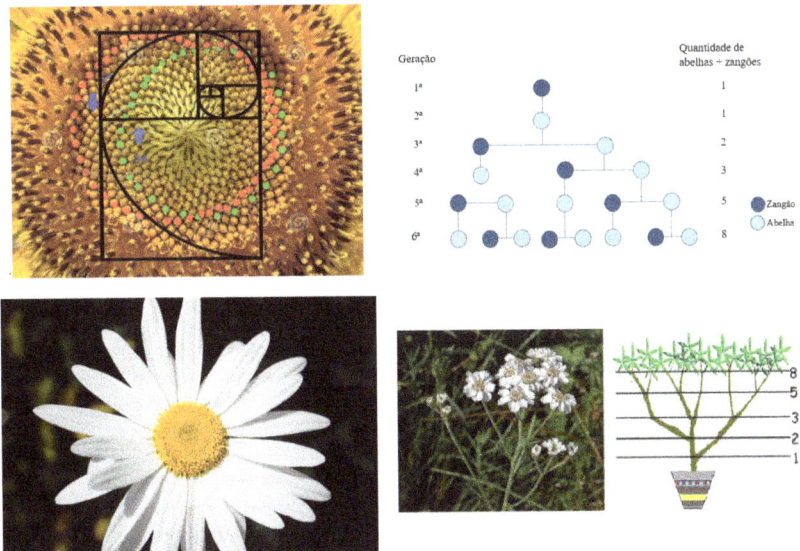

Fig. 4. Examples of figures presented to the seniors. Source: the internet.

3. Seniors' Engagement in a Landscape of Investigation

The first step in discussing the involvement of seniors inserted in a landscape of investigation concerns their *acceptance to participate*. According to Skovsmose (2011), this is the first (and perhaps most important) moment of a mathematical investigation in an educational context. The analysis of the data produced in this study indicates that the seniors accepted both proposals. For example, as soon as the team introduced the first scenario, even though they did not find a common characteristic among the images presented, the seniors tried to *find/perceive* regularities by asking questions, and they *took a stand, thinking aloud*.

Ana tried to identify geometric figures in common, Pedro observed that the figures were centralised and somewhat symmetrical, and Lúcia and Marina said that the curves were the characteristic the figures had in common. Lúcia *identified* a square, remembering a previous meeting where the group discussed magic squares, and asked whether the activity would be about curves. This finding was confirmed when the team showed them the same images, but this time with golden rectangles and spirals inserted, and the seniors *identified* the insertions and made statements such as: "*A proportional division was made.* [...] *It's like a shape...* [...] *It's like a template*" (Pedro), and "*the smallest part [of the spiral] is distributed in the whole*" (Sandra).

In the second landscape of investigation, the team considered that the seniors agreed to join the activity because of the warm discussions that emerged when they explored the rabbit problem, in which they tried not only to understand it but also to establish an action plan. At first, those discussions caused the group to fragment, because Joana was so convinced that her conjectures were correct that she decided to solve the problem by herself, while the other seniors remained in groups, working collaboratively.

Despite the participants' acceptance of the invitation to join the landscape of investigation, some seemed to feel unmotivated to continue exploring the situation. The reading of the text about the use of golden ratio in the Renaissance, and the discussion on questions and assumptions raised at the beginning of the golden ratio landscape of investigation, proved to be motivating for the introduction of the measurement of seniors' bodies. However, Ana, for instance, was not interested in the subject during this first stage. She engaged in the activity only when we started the actual body measurement. She began calculating in order to compare the measurements and ratios made with the golden ratio. At this stage, the project team measured some seniors, while some of the seniors measured each other. What we want to draw attention to here is the importance of the variation of activities in a landscape of investigation addressed to seniors. Invitations for certain activities may not be accepted by all participants, as happened with Ana, who became involved only when the practical activity did begin.

Besides, Lima (2015) states that the organisers of the activities must prepare for unexpected situations, such as the seniors' lack of interest or criticism toward the activity. In this way, this study reiterated, based on Julio and Oliveira (2018), that more than preparing an activity, it is paramount to prepare *for* it, thinking about possibilities of producing meanings, putting oneself in the place of the others (the seniors), and being open to interaction.

At different moments in both landscapes of investigation, the seniors *asked questions*. An example was in the golden ratio landscape of investigation, in which the reading made it possible to raise questions such as: "*Is there any relationship between the lower and upper limbs?*"

(Marina), and *"does this work for everyone?"* (Sandra). Glória commented that certain proportions might be related to European people, but that for other ethnic groups, such as the Chinese or Japanese, they would not be valid. Another example happened in the Fibonacci sequence scenario, when Lúcia commented that a sequence would be constructed, as she *realised* what would happen in the first months of the rabbit table. Regarding the elaboration of strategies in the Fibonacci sequence landscape of investigation, the very statement of the rabbit problem caused many disagreements between the seniors. It really was a challenge for them, and it was necessary to carry out a long process of negotiating strategies.

Alrø and Skovsmose (2003) highlight the importance of *evaluating* the perspective of the participants during the investigative process in a landscape of investigation, to understand their points of view about the problem and reach a common purpose. In the case of the Fibonacci sequence landscape of investigation, the rabbit problem was characterised by being a fictitious situation about the rabbits' breeding, which the seniors found difficult to detach from a real situation. Thus, the team needed to intervene in the discussion to clarify this issue, and to discuss in the group the objective of the problem, *restructuring it* from a perspective that was common to all—for example, considering the couple of rabbits from the first month as a young couple.

The participants were offered a large table which they should fill in together. They chose to remain seated, and as they completed the column of months, the person closest to the row referring to a given month filled it with young and adult rabbits (Fig. 5). Thus, everyone was able to help to construct the table, following what was being done, while some filled in their own tables at the same time.

Fig. 5. Setting up the table. Source: authors' collection, 2018.

As much as a table-filling strategy had already been established with the seniors, some of them were still having difficulties about what they should do. Thus, a break was needed to discuss the strategy in use with the group again. Only after all agreed and understood, were they able to continue. The strategy the seniors used to distribute rabbits each month was always to count the number of rabbits from the previous month that would be adults the next month and distribute them. Then, they checked the number of rabbit couples from the previous month that could have young couples next month, and distributed them to their respective parents. For example, in the eighth month, there were twenty-one rabbit couples: thirteen adult couples, and eight young couples (Fig. 6). Thus, to complete the ninth month, first, they placed twenty-one couples of adult rabbits and, as in the eighth month they had thirteen adult couples, only these couples would generate new bunnies in the ninth month so, the participants placed thirteen young couples so that each couple was beneath an adult couple. Thus, in the ninth month, thirty-four couples had been placed on the table.

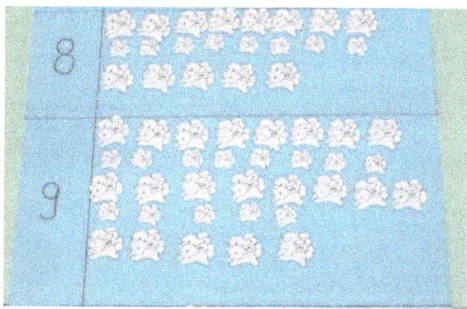

Fig. 6. Eighth and ninth month of the Table. Source: authors' collection, 2018.

Joana was working on her own during this task. When she saw that she had made miscalculations, she returned to the group, following the setting up of the table and helping to complete the tenth-month row. Even when filling in the table, Joana used a different strategy than the other seniors: she counted how many couples they had in each row in the previous month and distributed them in the following month. After she did this for all rows, she analysed which rows had adult couples in the previous month, and distributed the young couples in the following month. For example, in the ninth month, there were thirty-four couples of rabbits: twenty-one adult couples and thirteen young couples (Fig.

7). To complete the tenth month, Joana took the same amount of couples of rabbits as she had in the first row of the ninth month and placed them in the tenth month. As these eight couples were already adults in the ninth month, they would generate young couples in the tenth month. Therefore, Joana placed eight young couples corresponding to each adult couple above in the second row. After that, she took the same number of couples of rabbits from the second row of the ninth month and placed them in the third row of the tenth month. Since those couples were young in the ninth month, they would not generate young couples in the tenth month. Joana followed the same procedure for all the rows until she finished the tenth month.

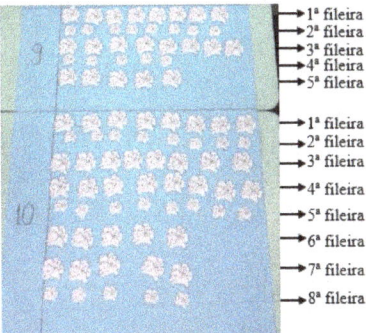

Fig. 7. Ninth and tenth month of the table. Source: authors' collection, 2018.

In the beginning, the other participants did not understand Joana's strategy, and, when asked, Joana explained it until everyone understood. Some women commented that *"changing the order of factors does not change the product"*. We noticed that this sharing encouraged the seniors to discuss the topic during the development of the activity. As already mentioned, in a landscape of investigation, the investigation process often becomes more important than the result itself (Skovsmose, 2011).

As in Lima et al. (2019), this study showed that the engagement of the seniors with landscapes of investigation allows them to develop creativity through the exposure of ideas and sharing of conclusions, leading to the emergence of different solutions to a problem and production of knowledge. In both meetings, the seniors *interacted/ collaborated actively* (Fig. 5). They decided in the group about the strategies for filling in the table by presenting different arguments, and they worked collaboratively to fill the table and discuss the number of rabbits. In Figure 8, which illustrates the measurement of bodies, we noticed a more significant interaction between the seniors and the project team.

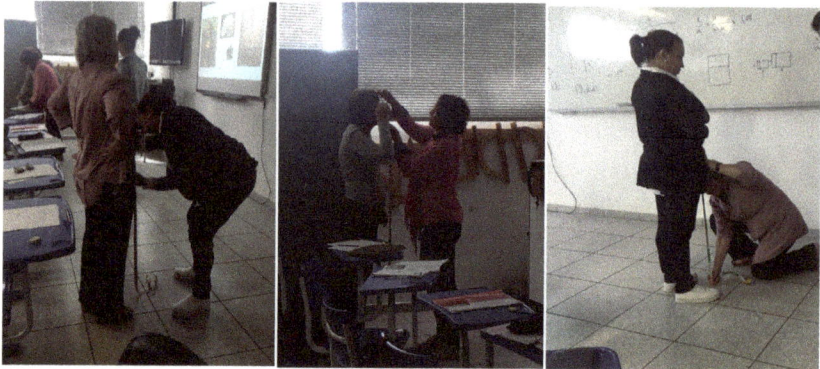

Fig. 8. Seniors taking measurements. Source: authors' collection, 2018.

Also, as in Lima et al. (2019), the analysis indicated that the seniors' engagement in a landscape of investigation fostered the *production of mathematical knowledge*. For example, when the seniors drew the golden rectangle with a ruler and a compass in the first activity, they did not find it hard to use them to construct the rectangle nor to calculate the golden ratio, using two different strategies: while some seniors calculated the ratio between the measurement of the larger side (AE) and the smaller side (AD), others calculated the ratio between the measurement of segments AB and BE, as shown in Figure 9. Then, each participant shared how they calculated, easing the interaction with each other, and showing they understood the subject.

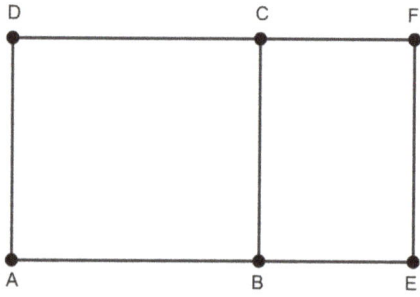

Fig. 9. Golden rectangle. Source: authors' collection, 2018.

In the Fibonacci sequence landscape of investigation, after the seniors filled out the cardboard table, the team proposed that they filled in their own tables. To conclude the investigation, we discussed the possible relationships between the numbers of the sequence formed in the column that represented the total of rabbits. In this way, the seniors could share their ideas, reasoning, and justification about the topic. The team that conducted the activity was in the background, trying to foster dialogue amongst the participants. At the beginning of the discussion, Sandra, Selma and Glória reported that they had already *realised* that each term, starting with the third, was obtained by adding the two previous terms (Fig. 10). They also found that this relationship occurred in the other columns (adult couples and young couples). The discussions they raised helped others to identify this pattern as well.

At this point, the research team chose to inform the seniors that the sequence formed was known as the Fibonacci sequence. The seniors continued to express thoughts and ideas about the relationships in the table out loud, realising that the sequence was formed in the other columns, starting in different rows. Sandra *noticed* another relationship between the elements in the table (Fig. 10), relating all the columns— that is, the sum of the total (of the first row) and the number of couples of young rabbits of the second row gave, as a result, the number of couples of adult rabbits of the third row; the sum of the total (of the second row) and the number of couples of young rabbits in the third row gave, as a result, the number of couples of adult rabbits in the fourth row, and so on. She raised this point and wanted to discuss it with the other seniors. Glória commented *"the table breeds like rabbits"*.

Month	Adult couples	Young couples	Total
1	0	1	1
2	1	0	1
3	1	1	2
4	2	1	3
5	3	2	5
6	5	3	8
7	8	5	13

Fig. 10. Table completed with the number of adult couples and young couples of rabbits and the total of couples of rabbits for each month. Source: authors' collection, 2018.

Sandra established regularities that the team had not noticed during the preparation of the activity. Other assumptions also surprised team members when carrying out the activities. For example, Joana tried to relate the total number of rabbits in the table to a percentage and questioned the existence of a growth graph of the situation explored. She also asked how we use this sequence and whether it is valid for other animals' breeding. We consider that, at that moment, the senior sought to *produce/relate* mathematical knowledge through the landscape of investigation. Also, the activities carried out provided the seniors with the ability to establish relationships between the activities carried out in the project. During the Fibonacci sequence landscape of investigation, the seniors had to fill a column in the individual table with the values of the ratio between the terms of the Fibonacci sequence. Using calculators, pencils, paper and mental calculation, right after the first divisions, they *realised* that the value was getting closer to the golden mean with which we had worked a few weeks before. Joana asked whether it would be possible to build a graph with those values. The team took the opportunity and presented a graph from Belini (2015) on the digital lab whiteboard with the values found in the divisions, aiming to corroborate the seniors' argument regarding obtaining the golden ratio (Fig. 11).

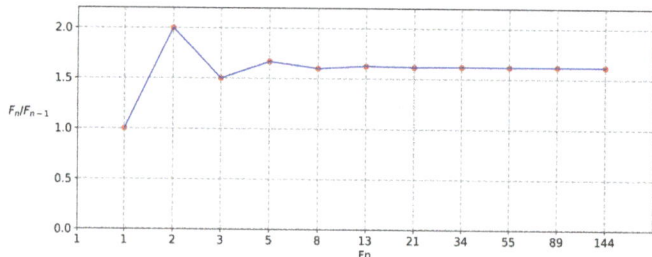

Fig. 11. Graph of the relationship between the Fibonacci sequence and the divisions. Source: Belini (2015, p. 39).

The analysis considered this situation to highlight that the seniors had understood the mathematical elements which the team had dealt with in that meeting and could relate them to what was being discussed during the Fibonacci sequence landscape of investigation. This corroborates that conducting mathematical investigations can provide opportunities for the seniors to use and rescue their mathematical knowledge, expanding their capacities and learning further (Lima, 2015; Lima et al., 2019). Another point to reinforce this perception came from one of Glória's statements: she was pleased to be able to talk with her family about those activities, citing as an example that she had told them that the golden ratio gives us the shape of a credit card. This can be read as an inclusive practice into family conversations for Glória, as she talked about subjects that they did not previously know about and they respected her knowledge, even those whom she considers as "experts in mathematics".

Seniors have a vast knowledge of the world. During activities developed throughout the Mathematics Conversations Project, they established relationships between mathematical approaches and people's daily practices and cultures. Thus, we consider that the open characteristic of landscapes of investigation linked to this knowledge of the world can foster the development of important moments of production of mathematical knowledge and different kinds of discussions. For the teacher or team who is in charge of activities, this can give rise to unforeseen moments, referred to in the literature as risk zones, characterised by the unpredictability and teachers losing control of the situation, which is even more likely to happen in a landscape of investigation (Penteado and Skovsmose, 2008; Silva and Penteado, 2013; Skovsmose, 2011; Penteado, 2001). These moments provided opportunities for the seniors and the executive team to discuss and undertake research into the topics. In other words, this research work verified that risks bring possibilities (Penteado and Skovsmose, 2008).

One of the characteristics of a landscape of investigation concerns the exploration of the different paths that may arise during the development of activities. As Skovsmose (2011) points out, the trails of the landscapes of investigation bring different modes of exploration. According to the author, there are moments to proceed slowly and cautiously, and others to throw yourself totally into the situation and see what happens. Those trails can often have an uncertain outcome. In the case of the landscapes of investigation covered in this chapter, there were several *discussions*—for example, about philosophy, aesthetics, technology and mathematical applicability. We consider these discussions as indicative of a senior's engagement in a landscape of investigation.

Philosophical discussions, for instance, were moments when the activities provided reflections concerning the very nature of mathematics. For example, during the golden ratio landscape of investigation, Sandra asked *"and the inventor of this [golden ratio], did he discover it?"*, which raised the question among the seniors whether the golden ratio was a human creation or a discovery. During this discussion, the strand of human discovery prevailed. In addition, this discussion reappeared at the end of the meeting, when images related to the golden mean were displayed, causing the belief among the seniors that "we discover" things done by a creator—in this case, God—to prevail. This is exemplified in Sandra's speech: *"I get angry at those atheists who think that everything came about by chance. How can they not believe in a superior being with such intelligence to create so many things that, for example, have the same proportion?"*

Ana initiated an example of an aesthetic discussion. In her words, *"this story of the golden ratio representing beauty is silly"*. For this senior woman, besides thinking that da Vinci's Mona Lisa is not beautiful, she did not use the golden ratio in her painting classes and did not think that its use could make the paintings more attractive. Other seniors also commented that the standard of beauty of the time could be quite different from today's because, if da Vinci made Mona Lisa using the golden ratio, it was probably considered beautiful at the time but, in their view, it would not be so today.

While carrying out the landscapes of investigation, the group *discussed the use of technology* both specifically and generally. As a specific aspect, the team observed that some of the seniors refused to use a calculator. This happened while calculating the proportions related to the golden ratio. One of the senior women explained that she does not use a calculator in her daily life, whereas another revealed that she wanted to do the mathematics by hand to remember the division's algorithm. In the Fibonacci sequence landscape of investigation, the ratios between the terms of the Fibonacci sequence required more complex calculations done with pencil and paper only. At this time, the seniors, except for one who refused to use the calculator, eventually discovered how easy it was to use such a device. According to Doll et al. (2016), the refusal to use the new technology, as occurred with the calculator, is an attitude expected from the seniors. They are generally not against the use of technology, but they do offer some resistance against it.

At the end of the meeting about the golden ratio, the team observed more general and critical comments on the use of technology. Selma commented on how technology interferes with the relationships between younger people, because nowadays, they prefer to chat through messaging. She also said that even straightforward attitudes such as saying good morning are undervalued today, showing a change in family upbringing. She noted that, in her teenage years, people needed to rely on their memory much more to store information such as phone numbers and birthdays, and that technology is replacing that today. According to Selma, this is a negative point of technology, as it can impact on the cognitive performance of young people concerning memorisation when they get old. During this activity, Selma also commented on the seniors' ability to discover and create things without today's technology resources, considering them more intelligent, and saying that *"all those seniors left us as knowledge, we only added to and developed"* (Selma).

Mathematics *applicability* in daily life was also a discussion the seniors brought into the activity. During the golden ratio landscape of investigation, it was necessary to bring more examples related to the determination of proportions between geometric shapes, which had not previously been thought of by the team. Thus, they also worked with the concept of similarity of shapes, which led Selma to ask: *"What is the applicability of it? How important is that in life?"* The research team answered that the golden ratio is used, for example, in some constructions, in arts and in the development of logos of some brands, because some people consider the golden rectangle a harmonious shape. In this sense, the landscape of investigation created routes for participants to go beyond mathematical concepts. They were able to talk about how mathematics can be seen in everyday life.

These discussions are exemplary in the sense of corroborating the claim of Skovsmose (2011) about the different paths that can arise in an investigative process. The project team took those discussions seriously, trying to incorporate them into activities to offer the seniors a chance to see mathematics differently and become more eager to learn it. They ended up showing interest, as the team could see. In one evaluation meeting, Pedro commented that he found the meeting very interesting, that it was additional knowledge for his life, and that he found it very practical. Sandra commented: *"It is additional knowledge [...]. I loved* talking about constants. [...] I'm thinking about enrolling in a mathematics undergraduate program in the next year; I was traumatised with mathematics because I failed the second grade, but I have been reading books since some years ago, and I got involved with mathematics"*. Simone, for example, pointed out: *"I even learned to understand mathematics in nature. [...] Culturally, it was very interesting, wonderful."* Glória said: *"I loved it, I learned a lot, but for me, as the oldest participant, it was too much information for a single class".*

4. Final Remarks

In this chapter we discussed the engagement of seniors in landscapes of investigation. This study reveals that the participation of seniors initially occurred after them accepting the invitation to get involved in the activities. It was possible to observe that, in some moments, the seniors explored problems and situations by formulating questions, developing tests and strategies, and defending ideas. The activities were pervaded by questions and discussions, while the seniors were seeking to understand the subjects and the mathematical concepts involved, showing their interest. Another essential aspect of engagement was the interaction amongst them in carrying out the activities and working with the project team's members.

As evidenced throughout this chapter, the landscapes of investigation built to deal with the golden ratio and Fibonacci sequence opened up possibilities for the seniors in discussing various topics. They produced mathematical knowledge. This activity can be seen as an opportunity to acquire a more positive view of mathematics and to learn new things—an essential aspect of the life of a senior. During the activities, the project team tried to create an environment where the participants felt free to express themselves on any subject, related (or not) to the theme. The results discussed in this chapter are aligned with studies such as Lima (2015), Lima et al. (2019), Grossi (2014) and Scagion (2018), since this type of educational work can foster seniors' self-esteem and provide opportunities for intergenerational dialogue by sharing their knowledge and experiences. In addition, the attitudes of the participants regarding asking questions, taking notes and carrying out some of the activities in their homes also evidences their commitment and willingness to learn.

References

Alrø, H. & Skovsmose, O. (2003). *Dialogue and learning in mathematics education: Intention, reflection, critique.* Kluwer Academic Publishers.

Belini, M. M. (2015). *A razão áurea e a sequência de Fibonacci.* Master's thesis. Universidade de São Paulo (USP).

Doll, J. (2008). Educação e envelhecimento: fundamentos e perspectivas. A terceira Idade: *Estudos sobre Envelhecimento, 19*(43), 7–26.

Doll, J., Machado, L. R. & Cachioni, M. (2016). O idoso e as novas tecnologias. In E. V. d. Freitas & L. Py (Eds), *Tratado de geriatria e gerontologia* (pp. 1654–1663). Guanabara Koogan.

Freire, P. (2016). *Pedagogia do oprimido* (60th edition). Paz e Terra.

Garcia, V. C., Serres, F. F., Magro, J. Z. & Azevedo, B. d. (sn). *O número de ouro como instrumento de aprendizagem significativa no estudo dos números irracionais.* Apostila de graduação. Universidade Federal do Rio Grande do Sul.

Grossi, F. C. D. P. (2014). *Os diferentes "lugares" que a escola, a leitura, a escrita e a aula de matemática têm na vida dos alunos que estão na terceira idade.* Master's thesis. Universidade Federal de São João del Rey.

IBGE. (2019). *Instituto Brasileiro de Geografia e Estatística. Síntese de Indicadores sociais: uma análise das condições de vida da população brasileira—2019* (Vol. 40). Coordenação de População e Indicadores Sociais.

Julio, R. S. & Oliveira, V. C. A. (2018). Estranhamento e descentramento na prática de formação de professores de Matemática. *Boletim GEPEM, 72,* 112–123.

Julio, R. S. & Silva, G. H. G. d. (2019). Educação Matemática, inclusão social e pessoas idosas: uma análise do projeto conversas matemáticas no âmbito do Programa Universidade Aberta à Pessoa Idosa. *Educação Matemática em Revista, 24*(64), 52–70.

Law n. 10.741 of October 2003 (2003).

Law n. 13.005 of June 2014 (2014).

Leopoldino, K. S. M. (2016). *Sequências de Fibonacci e a Razão Áurea: aplicações no ensino básico.* Master's thesis. Universidade Federal do Rio Grande do Norte.

Lima, L. F. d. (2015). *Conversas sobre matemática com pessoas idosas viabilizadas por uma ação de extensão universitária.* Doctoral dissertation. Universidade Estadual Paulista (Unesp).

Lima, L. F. d., Penteado, M. G. & Silva, G. H. G. d. (2019). Há sempre o que ensinar, há sempre o que aprender: como e por que educação matemática na terceira idade? *Boletim de Educação Matemática (BOLEMA), 33*(65), 1331–1356.

Lívio, M. (2008). *Razão Áurea: a história de fi, um número surpreendente*. Record.

Moscovici, S. (2005). *Representações sociais: investigações em psicologia social*. Editora Vozes.

Penteado, M. G. (2001). Computer-based learning environments: Risks and uncertainties for teacher. *Ways of Knowing Journal, 1*(2), 23–35.

Penteado, M. G. & Skovsmose, O. (2008). Riscos trazem possibilidades. In O. Skovsmose (Ed.), *Desafios da reflexão em Educação Matemática Crítica* (pp. 41–50). Papirus.

Queiroz, R. M. (2007). Razão áurea: a beleza de uma razão surpreendente. In U. E. d. Londrina (Ed.), *Professional development program*. Universidade Estadual de Londrina (UEL).

Scagion, M. P. (2018). *Representações sociais de pessoas idosas sobre matemática*. Master's thesis. Universidade Estadual Paulista (Unesp).

Scortegagna, P. A. (2010). *Políticas públicas e a educação para a terceira idade: contornos, controvérsias e possibilidades*. Master's thesis. Universidade Estadual de Ponta Grossa.

Silva, G. H. G. & Penteado, M. G. (2013). Geometria dinâmica na sala de aula: o desenvolvimento do futuro professor de matemática diante da imprevisibilidade. *Ciência & Educação, 19*(2), 13.

Silva, N. d. (2020). *Educação Matemática a partir de um projeto de extensão direcionado a pessoas idosas: Contribuições para a formação inicial de professores de matemática*. Master's thesis. Universidade Federal de Alfenas.

Silva, R. N. d., Silva, G. H. G. d. & Julio, R. S. (2019). Educação matemática e atividades com pessoas idosas. *Revista Debate em Educação, 9*(1), 560–587.

Skovsmose, O. (2000). Cenários para investigação. *Boletim de Educação Matemática (BOLEMA)* (14), 66–91.

Skovsmose, O. (2011). *An invitation to critical mathematics education*. Sense Publishers.

15. The Investigative Approach to Talking about Inclusion in Mathematics Teacher Education

Denner Dias Barros

Looking at the range of diversity in the classroom is a necessary task for the teacher who seeks to teach from an inclusive perspective. More and more educational legislation has been authorised so that all students can have access to quality education. The use of landscapes of investigation in mathematics teacher education could become a practice that helps prospective teachers to reflect on future practices by taking into consideration the diversity of school contexts. In this chapter, reflections are made on the practices of two research projects, one for a doctorate and another for a Master's degree. Working with investigative projects allowed prospective teachers to think about practices that take into account the particularities of this diversity. Furthermore, the dialogic perspective is highlighted as being essential in this process, as well as the ideas and visions of the prospective teachers invited to participate in their own professional education. It is hoped that this chapter can inspire prospective teachers to develop critical activities that allow for the participation of all students.

In Brazil, the end of the twentieth century and the beginning of the twenty-first brought educational policies aimed at greater democratisation of access and a quality education for all. These policies propose that groups which were historically excluded from the school space may have the right to quality education, and that educational institutions reorganise themselves in order to meet the specific needs of all students. This initiative can be understood as a paradigmatic shift, as it indicates that it is not the student who must adapt to a standard imposed by the school, but the school that must adapt itself to become a more receptive space for differences.

From the perspective of inclusive education, the National Policy on Special Education (2008) deals with the necessary changes to make schools more inclusive. It reinforces the need for greater investments in infrastructure, acquisition of materials and accessibility, hiring of specialist professionals—such as specialised educational service teachers and Brazilian Sign Language (Libras) interpreters—and changes in the initial and continuing education of teachers.

In Brazil, deaf communities use the sign language called Libras, which is understood as the official means of communication for this population. Its use was promulgated by a law from 2002. Regarding deaf education, a decree from 2005 represented a great advance for this group. It established bilingual education, where deaf people have the right to have Libras as their first language and Portuguese in written form as their second.

A law passed in 2002 establishes the National Curriculum Guidelines for the Training of Teachers in Basic Education. This legislation requires that higher education institutions provide a curriculum structured around a teacher training programme focussed on the differences and plurality of the classroom, for which, in addition, all students with disabilities must also be considered.

However, we must take into account that the teacher's pedagogical practice is comprised of multiple dimensions. According to Tardif (2002), looking at teacher training in this way makes it possible to analyse what a prospective teacher should be taught in initial training, since from there he or she builds up what can be called "teaching knowledge". Such knowledge becomes further developed throughout the teacher's subsequent practice, but it is in initial teacher training that the prospective teacher will have contact with a more systematised and explicit form of "teaching knowledge".

From the perspective of inclusive education, the classroom is seen as a space for unifying difference. In teacher education, it is necessary to articulate the development of theoretical and practical experiences, by taking into consideration the uniqueness of all students and the different working conditions of teachers.

In discussions about basic education within teacher education, we take into account the values that explain human life, and social and cultural issues (Silva Júnior, 2010). This is an essential theme in teacher education, so that these professionals are aware of and respect students' differences, whether they are related to ethnicity, gender, social condition, disability and others.

Addressing pupil diversity in teacher education by keeping such a vast range of characteristics of students in mind is still a challenge. When teachers finish their formal training and start working, they often feel insecure and unprepared to deal with everyday school life. According to Tardif and Raymond (2000), the beginning of a career as a teacher is full of doubts, as it confronts the prospective teachers' initial expectations with the reality of school life.

In this context, this text presents a reflection on the training of mathematics teachers for coping with diversities. In particular, research practices are defended as a way to address these issues with prospective teachers.

1. Investigative Experiences in the Training of Mathematics Teachers to Reflect on Inclusive Education

Research by Cintra (2014) and Barros (2017), linked to Épura—a research group focussing on mathematics education and inclusion at the Universidade Estadual Paulista, Unesp, campus of Rio Claro—reflects on investigative practices in the training of mathematics teachers that were carried out with the aim of discussing inclusive education and the importance of valuing differences.

Cintra investigated what understandings could be produced from working with projects addressing inclusive education in the initial training of mathematics teachers. The research was carried out in two courses (Development of Project 1 and Development of Project 2) in a mathematics teacher education programme at the public university in the state of Minas Gerais. Prospective teachers were categorised as researchers, and in this way had the opportunity to adopt a reflective and investigative approach to their professional learning. Consequently, they came to question their previous understandings about school inclusion.

In developing an investigative project, students had the opportunity to exercise their autonomy and work collaboratively in groups. The prospective mathematics teachers first undertook a theoretical study on the theme of school inclusion. In this exercise, they developed material that later guided their own production of data, which they analysed and presented to other prospective teachers. The development of the project was guided by Cintra.

Cintra's students collectively decided that they would investigate the theme of inclusion. The topics related to inclusion that were chosen to be studied in depth were:

- The use of technologies to support students with visual and/or auditory impairments;
- The process of transcription of mathematics books;
- Designing and using mathematical activities in a classroom with both deaf and hearing students in a regular school;
- Application of mathematics activities for blind students.

According to Cintra, the elaboration of research projects on inclusion made it possible for the prospective mathematics teachers to reflect on the presence of students with disabilities at the school, and on the need to respect their differences. The prospective teachers came to recognise the potential of these students. Cintra points out that working with research projects was a challenge that generated expectations, opportunities and challenges for both the professor and the students.

> ... asseguramos que trabalhar com projetos na formação inicial é uma metodologia que privilegia a investigação e a produção de conhecimento. No caso específico desta tese, auxiliou os futuros professores na compreensão de situações complexas de ensinar e aprender para a diversidade, desenvolvendo uma atitude crítica em relação à Educação Especial. [... we ensure that working with projects in initial training is a methodology that favours research and knowledge production. In the specific case of this thesis, it helped prospective teachers to understand complex situations of teaching and learning for diversity, developing a critical attitude towards special education] (Cintra, 2014, p. 129, my translation).

In my Master's thesis, *Initial Training of Mathematics Teachers From the Perspective of Inclusive Education: Contributions From the Libras Course* (Barros, 2017), I looked at possibilities for discussing inclusion in a Libras course for prospective mathematics teachers at a public university in the state of São Paulo. This course is a course in the graduate programme, where prospective teachers in mathematics can learn Libras, as well as learning about being a deaf person. This course also presents possibilities for teaching mathematics classes for deaf and hearing people, and possibilities for using inclusive methodologies.

In order to collaborate with this bilingual education for the deaf, the decree from 2005 added the obligation to offer a Libras course within teacher training and speech therapy courses. However, the legislation did not provide guidelines on how to implement and organise the courses. Whilst this was a great gain for teacher education, it was also a challenge, since there are no specific guidelines that guarantee a course that addresses the need to consider inclusion and diversity in the initial training of these professionals, although there are some indications about the importance of these themes in the training of teachers. Therefore, higher education institutions have been reflecting on how to organise this education in the best way.

My work with the Libras course was aimed at understanding how the course was constituted. The Libras course was not only meant to address the teaching of the language, but also had to make sure that the prospective mathematics teachers who attended the course had the opportunity to reflect on the concept of inclusion and possible inclusive practices that could become part of their training on working with differences in the classroom.

In this process, prospective teachers were able to investigate the use of technology resources and methodologies that enhance the participation of students with disabilities in maths classes, as well as the establishment of partnerships seeking collaborative work between teachers and professionals in special education.

Just like in Cintra's research, the Libras course was developed using a methodology involving projects. The project engaged the prospective teachers in imagining a classroom that would contain at least one student with a disability. The prospective teacher should choose a piece of mathematical content and discuss how this project could be developed in that context with an inclusive perspective. The work within these projects was used in the Libras course to ensure that the prospective teachers could think about how to teach mathematics, considering the differences in the classroom and seeking a quality education for all.

I interviewed the teachers of the course, who highlighted that they thought that giving prospective teachers the opportunity to work with projects offered appropriate conditions for them to recognise the importance of differences and the possibilities for inclusive practices.

Facing reality as a problem to be addressed helps to overcome the view that social problems are not a collective responsibility. Since the future is seen in this way, students can feel like agents of transformation in society. Working with projects can help in this regard, as:

> ... favorece a concepção de realidade como um fato problemático que é preciso resolver e responde ao princípio de integração e de totalidade, e que leva ao ensino globalizado, isto é, não existem cadeiras isoladas, porque os projetos incluem todos os aspectos da aprendizagem: leitura, escrita, cálculo, expressão gráfica, etc. [It favours the conception of reality as a problematic fact that needs to be resolved and responds to the principle of integration and totality, and that leads to globalised teaching, that is, there are no isolated chairs, because the projects include all aspects of learning: reading, writing, calculation, graphic expression, etc.] (Zabala, 2002, p. 205, my translation).

Reflecting on this perspective of Zabala, we can understand that projects naturally have an interdisciplinary characteristic, and a project developed in the context of mathematics education can mobilise knowledge from several other areas, as well as an investigation into the diversity of the classroom—that is, reflections on the very idea of inclusive education.

2. Landscapes of Investigation as a Possibility in Teacher Training When Acknowledging Differences

When considering Cintra's and my research, a question arises: is it possible to approximate the practices analysed in these research projects with the proposal of landscapes of investigation? I will consider this question in the following.

In a landscape of investigation, both teachers and students are understood as people who teach and learn. A landscape of investigation is a way of looking at the world, thinking about relationships, putting yourself in an investigative posture and feeling part of society.

Furthermore, landscapes of investigation allow activities to be conducted in different ways, respecting different learning processes.

> Such a landscape provides an environment for teaching-learning activities. While sequences of exercises, so characteristics for the school mathematics tradition, establish a one-way route through the curriculum, the possible routes through a landscape of investigation are not well-defined. A landscape can be explored in different manners and through different routes. Sometimes one must proceed slowly and carefully, sometimes one can jump around and make bold guesses (Skovsmose, 2011, p. 31).

In working with a landscape of investigation, the first step is to invite students to participate in the activity (Skovsmose, 2001). Problems investigated in an investigative setting need to make sense to the participants so that prospective teachers can mobilise themselves in the search for solutions. Cintra was attentive to this issue, allowing her students to choose the investigated themes. In the course that I investigated, the teachers provided the students with the possibility of outlining a school reality. They wanted to later reflect on how it would be possible to develop an inclusive practice in that context.

Furthermore, making this invitation does not only concern the prospective teachers accepting or refusing to participate in the activity, but it is a possibility for the teacher-trainers to look at their own practice. As a result, they can then decide whether it is necessary, when reflecting on how the prospective teachers received the proposal, to reformulate the invitation and the activity such that they get closer to the life, anxieties, desires and problems of the prospective teachers.

In an investigative setting, dialogue is essential. For Freire (1987), dialogue is fundamental to freedom in the act of learning. Dialogue concerns interpersonal relationships and engagement. Freire defines dialogue as a meeting between people with the objective of "naming the world" in a relationship of respect with humility.

In Cintra (2014), we see the valorisation of group work and the partnership established between the teacher of the course and prospective teachers, in a search to understand, investigate and share information and opinions. In Barros (2017), the teachers who taught the course report on the importance of sharing their practices with prospective teachers, and also of inviting deaf people to join the Libras course and talk about deaf people's experiences of school life.

When reflecting on the formation of teacher knowledge, Tardif (2002) says that prospective teachers also build their knowledge by listening, observing and dialoguing with their own teachers. Thus, experiences of teaching practice can be shared and situations experienced can be used to foster discussions in initial training.

Alrø and Skovsmose (2004) point out that the traditional mathematics classroom often adopts a communication pattern which they refer to as "bureaucratic absolutism". Within this pattern, the teachers position themselves as the holders of knowledge and, in an authoritarian way, validate what is stated by the students, assuming that mathematics is constituted by absolute and unquestionable truths. Dialogue, in turn, appears in contrast to this posture. For the authors, there are some dialogical acts that constitute what they call "investigative cooperation", where instead of a vertical and authoritarian relationship, teachers and students place themselves in a horizontal and dialogical relationship, where they can establish contact, perceive, recognise, position, think out loud, reformulate, challenge and evaluate.

In a course that seeks to conduct discussions on mathematics education and inclusion in teacher education, it is also important to establish a horizontal and dialogical relationship, so that both the teacher-trainers and the prospective teachers can reflect on possibilities of inclusive practices and of valuing differences.

Inspired by Penteado (2001), Skovsmose (2001) reflects on how working from a perspective of landscapes of investigation is, for teachers, equivalent to placing themselves in a risk zone. A dialogic relationship is marked by unpredictability. However, being in a risk zone also means being open to new possibilities, so that reflections that were not previously established by the professor can emerge and contribute to the understanding of what is being investigated.

Cintra (2014, p. 132) highlights that, by adopting an investigative approach, prospective teachers could learn about themselves and their subsequent teaching practice, and that the researchers could also learn throughout this process by placing themselves in a position of researching alongside the students. The teacher-educators that I interviewed (see Barros, 2017) pointed out that the course itself adapted each year, as it consisted of different students. Furthermore, an investigative approach allowed the teacher-educators to reflect on their practice and on ways to align their perspectives with the real needs of these teachers in training.

Engaging in investigative practices within landscapes of investigation can enable prospective mathematics teachers to reflect on their future practices and value differences. Bringing landscapes of investigation into mathematics teacher education can establish conditions for exchanges of ideas, opinions, uncertainties and stimulated reflection on possible practices for inclusion.

3. Some Considerations

Taking into account the assumptions of inclusive education, we see that prospective mathematics teachers can be placed in an investigative stance on pupil difference. Teaching from an inclusive perspective presupposes a dialogical environment. After all, students are different and learning processes are different. Casting this gaze on students and engaging in the process of dialogue is also essential for differences to be recognised and valued.

Furthermore, the investigation cannot be an imposed thing, but a process in which the students are invited to participate. An essential first step for a critical and questioning education is to raise awareness about awareness itself, so that students can see the importance of a reflective education that questions reality (Freire, 2011). This issue of involvement with the investigated problem was highlighted in both analysed studies, when the prospective teachers had autonomy in choosing the topic and in the design of the steps of the investigation.

Education must be given new meaning through practices that promote the development of critical thinking and the appreciation of differences in the classroom. For this, other educational possibilities need to be presented; ones that question the idea that there is a single classroom model and that all students are identical. We understand that the research project proposals that are studied in the research by Cintra and by me were constituted as landscapes of investigation with this purpose.

There are many factors that interfere in the search for school inclusion; after all, public policies, accessibility actions and the availability of resources and services are necessary. However, the development of a positive view of differences by mathematics teachers is essential for establishing a more inclusive school.

References

Alrø, H. & Skovsmose, O. (2004). *Dialogue and learning in mathematics education: Intention, reflection, critique*. Kluwer Academic Publishers.

Barros, D. D. (2017). *Formação inicial de professores de matemática na perspectiva da educação inclusiva: Contribuições da disciplina de Libras*. Master's thesis. Universidade Estadual Paulista (Unesp).

Brasil. (2001). *Ministério da Educação. Secretaria de Educação Especial. Diretrizes Nacionais para a Educação Especial na Educação Básica*. MEC/SEESP.

Brasil. (2008). *Ministério da Educação. Secretaria da Educação Especial. Política Nacional de Educação Especial na perspectiva da Educação Inclusiva*. MEC/SEESP.

Cintra, V. P. (2014). *Trabalho com projetos na formação inicial de professores de matemática na perspectiva da educação inclusiva*. Doctoral dissertation. Universidade Estadual Paulista (Unesp).

Freire, P. (1987). *Pedagogia do oprimido* (17th edition). Paz e Terra.

Freire, P. (2011). *Educação como prática de liberdade* (14th edition). Paz e Terra.

Penteado, M. G. (2001). Computer-based learning environments: Risks and uncertainties for teachers. *Ways of Knowing, 1*(2), 23–35.

Rosa, E. A. C. (2014). *Professores que ensinam matemática e a inclusão escolar: algumas apreensões*. Master's thesis. Universidade Estadual Paulista (Unesp).

Silva Júnior, C. A. (2010). *Fortalecimento das políticas de valorização docente: proposição de novos formatos para cursos de licenciatura para o estado da Bahia*. UNESCO/CAPES. (Relatório. Documento interno).

Skovsmose, O. (2001). Landscapes of investigation. *ZDM—Mathematics Education, 33*(4), 123–132. Reprinted as Chapter 1 in Skovsmose, O. (2014), Critique as uncertainty (pp. 3–20). Information Age Publishing.

Skovsmose, O. (2011). *An invitation to critical mathematics education*. Sense Publishers.

Tardif, M. (2002). *Saberes docentes e formação profissional*. Editora Vozes.

Tardif, M. & Raymond, D. (2000). Saberes, tempo e aprendizagem do trabalho no magistério. *Educação e Sociedade, 21*(73), 209–244.

Zabala, A. (2002). *Enfoque globalizador e pensamento complexo*. ArtMed.

16. Opening an Exercise: Prospective Mathematics Teachers Entering into Landscapes of Investigation

Raquel Milani

This text presents two possibilities for transforming exercises of the school mathematics tradition into investigative activities related to landscapes of investigation located in the context of critical mathematics education (CME). This process of transformation was based on what Skovsmose calls "opening an exercise". In order to open an exercise with reference to pure mathematics, one can explore the theme proposed in the exercise by going in different directions. In order to open an exercise with reference to semi-reality and reality, I have proposed the exploration of comments produced by students as they work through the exercise. A task related to such transformation was developed for prospective mathematics teachers in order to reflect on mathemacy, help students to make mathematical discoveries, and potentially cause changes in traditional teaching practices. The result of the transformation process carried out by the prospective teachers in the course indicates that the investigative activity should have a strong appeal to reality and semi-reality, besides mobilising other actions that are not developed when one solves an exercise. Due to the characteristics of the task, a new reference emerged: an imagined reality. The task and reflections represent the first step in the process of these prospective teachers entering CME.

1. Milieus of Learning in Mathematics Classes

In a Brazilian public university, I taught a course about methodologies for mathematics teaching. The group of prospective mathematics teachers had already studied, in other disciplines, mathematical subjects such as integral and differential calculus, algebra and geometry. They were about to start their supervised teaching practices. With the group of prospective teachers, I discussed some notions of critical mathematics education (CME) according to Skovsmose's paper about *milieus of learning*. For the vast majority of them, that was the first time they had heard of the subject. These notions will be presented in this text and form the basis of a task designed for the prospective teachers to transform traditional exercises into investigative activities.

Mathematics classes can be organised in different ways, either by the type of activity developed by the students, or by the kind of communication between the teacher and the students. In this sense, Skovsmose (2000) presents this diversity referring to milieus of learning, arranged according to a matrix which was presented properly in the first chapter of this book.

This table presents six milieus of learning located in the *paradigm of exercise* or in *landscapes of investigation*, with different references: *pure mathematics, semi-reality* and *reality* (Skovsmose, 2000). The first reference is about activities whose context is strictly mathematical. The reference to a semi-reality is related to a constructed reality, and not to an observed reality—thus the information presented in an activity refers to situations that *might* occur or exist. In activities with reference to reality, students and teachers work with real-life situations. Examples of each learning environment will be presented in the following section.

Skovsmose characterises the school mathematics tradition as one which is particularly set in the paradigm of exercise. In this context, bureaucratic absolutism appears "to draw on unlimited resources for stating in absolute terms what is right and what is wrong" (Alrø and Skovsmose, 2004, p. 26). In these environments, the goal is to train in a technique, and memorise concepts and procedures by repetition. Students usually face the blackboard. Teachers present mathematical techniques, ideas and examples, and then the students solve some exercises, selected from the textbooks, that have only one answer (Skovsmose, 2000, 2001).

In the paradigm of exercise, it is possible to find patterns of communication where the teacher usually asks questions which have a unique answer that is known by them in advance. The students, in turn, try to guess what they want in response. The teacher evaluates their answers as right or wrong, representing the authority in the classroom. The teacher's task is to explain the algorithms and correct the students' errors.

On the other hand, aiming for critical learning, Skovsmose (2000) presents landscapes of investigation with activities that can help the students to discover mathematical facts, as well as reflecting, understanding and making decisions based on facts of reality. In this context, students' participation is active and, when working in groups, they develop dialogic acts with their peers and teacher, which is important for learning.

If students are actively engaged in an investigative activity, it is because they have accepted the teacher's invitation to develop the investigation. There is an intention and an attitude of curiosity that moves the participants. A landscape of investigation is established to provide meaning to the activities in which the students are participating (Skovsmose, 2011). They are invited to explore hypotheses and make discoveries. Once they are engaged in the activity, the teacher cannot anticipate what the students will discover in their investigation. This is because they can choose the path to be followed in the investigation and act according to their decisions. Some questions may be proposed by the teacher, but others may arise during the activity, leading to new possibilities of investigation. An investigative activity, therefore, is characterised by a high degree of unpredictability. One does not look for genuine results in mathematics, but rather for the students to make their own discoveries (Skovsmose, 2011).

Critical learning of mathematics is related to *mathemacy* (Skovsmose, 2007, 2011), which has a proximal relationship with the notion of literacy as described by Paulo Freire, and that of matheracy by Ubiratan D'Ambrosio. Literacy involves more than competence in reading and writing. It also requires the ability to interpret social, cultural, political and economic situations (life-worlds) and try actively to change these situations (Skovsmose, 2011). In a similar way, besides requiring competence in handling mathematical techniques, "mathemacy can be seen as a way of reading the world in terms of numbers and figures, and of writing it as being open to change" (Skovsmose, 2011, p. 83). With mathemacy, one can critically read situations where mathematical concepts appear in an explicit or implicit way. These situations are called *mathematics in action* by Skovsmose. One can evaluate the power positions, the risks involved and possible changes related to those situations. With mathemacy, the student can change the way they look at the situation, and it could help them to change their usual actions in the face of that situation.

In the context of CME, Skovsmose's text from 2000 is widely used in Brazilian research on mathematics classes. Analysing the thematic edition of a Brazilian known journal (*Revista Paranaense de Educação Matemática*) on CME, Marcone and Milani (2020) state that 60% of the scientific papers refer directly or indirectly to Skovsmose's text. Regularly in my practice, when I discuss this text with school teachers and prospective teachers, the vast majority state that their teaching practices are located in cells (1) and (3) of the milieus of the learning matrix, and landscapes of investigation represent difficulties or challenges for them. If one wonders about changing the traditional teaching practices to assimilate them to landscapes of investigation, a first step could be starting from what the teachers already do in their practices—that is, activities based on the paradigm of exercises.

But how does one transform exercises into investigative activities? A possibility for doing so was presented by myself in the course of a mathematics teacher education programme with the aim of discussing possible changes in the school mathematics tradition with the prospective teachers. The intention was to generate discussions about how the teachers could promote activities focussed on the development of mathemacy and enable the students to make discoveries in the context of pure mathematics, characterising the activities of landscapes of investigation. The description of this work in the course is the focus of the next section.

2. Preparing the Environment to Open an Exercise

The context of the research was a final-year course of a mathematics teacher education programme. Until this moment, the prospective teachers had already attended both pedagogical and mathematical content courses. In a certain class, I discussed with them the milieus of learning presented by Skovsmose (2000). Out of a total of sixty prospective teachers, distributed in two classes, only two expressed knowing something about CME; the notion of landscapes of investigation was a novelty to the overwhelming majority of students. The reading and discussion of this text, therefore, represented the gateway to thinking about CME.

First, the prospective teachers read the text. The discussion in class was guided by a PowerPoint presentation which I had made, exploring at least one example from each milieu of learning. To address an exercise with reference to pure mathematics, I presented the following example: "For the functions and from to defined through $f(x) = 2x + 6$ and $g(x) = -2x + 5$, find the equation that defines the functions f^{-1} and g^{-1}." The goal of this exercise is for students to practise a technique to find the inverse of a linear function.

How does one transform that exercise into an activity related to landscapes of investigation? One possibility for that is presented by Skovsmose (2011), when the author introduces the action of *opening an exercise*, with reference to pure mathematics. The author himself asks: "What could it mean to open an exercise and try to enter a landscape of investigation through this opening?" (Skovsmose, 2011, p. 32). The author proposes a situation similar to the following one: "Let us consider two linear functions f and g from R to R defined through $f(x) = ax + b$ and $g(x) = cx + d$, where the parameters a, b, c, d are real values, a and c different from zero. What could we say about the intersection of the functions f and g? And of f and f^{-1}? And of f^{-1} and g^{-1}?"

Unlike the exercise presented earlier, there is no unique answer expected by the teacher for this investigative activity with reference to pure mathematics. Of course, to solve the above exercise, the students need to know how to find the equation of the inverse function, but the task does not stop there. They are invited to explore mathematical concepts and hypotheses: for example, cases of intersection and graphs of first-degree function. In this situation, three questions are presented, and it indicates that there are others that can be formulated by the teacher or the students, depending on their intentions with the activity. This example of opening an exercise to enter into a landscape of investigation shows that there are other possibilities of work with reference to the theme proposed in the exercise. But what about the references to semi-reality and reality? How can we make the transition from the paradigm of exercise to landscapes of investigation? What could it mean to open an exercise when the reference is to semi-reality or reality? Skovsmose (2000, 2011) does not directly address this issue. In my course with the prospective teachers, I presented a possibility.

As an example of an exercise referencing semi-reality, I presented the following: "John went to a green market to buy 5 kg of apples. If the price of each kilogram of apples was R$ 13.50, how much did John pay for the purchase?" This is a classic example in Brazilian textbooks that gives the context of a person at a fruit and vegetable market who buys a certain product at a certain cost, and the learner is asked to calculate the total amount spent by that person.

One does not know who John is, but it is possible that John from the exercise exists. One does not know if that market exists, but it is probable that there is some green market that sells apples at this price. It is possible that some John at some market buys 5 kg of apples at R$ 13.50 per kilogram. These possibilities make this exercise fit into the reference of a semi-reality. One possibility of transforming this exercise into an investigative activity could be to listen to the students' comments about the exercise while they are reading, listening and thinking about the information presented in the exercise. Some of these comments could be: "This apple is so expensive! What will John do with all these apples? No one ever buys exactly 5 kg of apples! And at the market near my house, I can find apples much cheaper than that! I wonder if you have to pay that price in all green markets. Why didn't he search for a better price somewhere else? How much did it cost the farmer to produce this apple? This apple must be made of gold!"

All these comments can be silenced by the teacher if the goal is to work only in the paradigm of exercise. "That is OK! Now, solve the exercise. How much did John pay for the apples?", the teacher may say. This stance assumed by the teacher is very usual in Brazilian mathematics classrooms. Questioning the data of the exercise and its context is not a frequent act, not even by the textbook's authors. However, starting from each one of those comments enunciated by the students, an investigative activity may emerge. For example, let us consider the sentence: "No one ever buys exactly 5 kg of apples!"

In a green market, when fruit, for example, is placed on the scale, it is difficult to get the exact quantity intended of the product (5 kg, as in the exercise above). When that quantity is a little larger than what was requested by the costumer (5.1 kg, for example), the seller usually rounds this amount down and charges for the amount requested by the costumer. With this situation in mind, the teacher and students may wonder: How much does the seller lose at the end of a working day by rounding down the quantity of the products sold? Is it really a loss? What are the benefits to the seller and the customer with such rounding? In São Paulo, there are many green markets spread across the city and the citizens often go there and buy their goods. Thus, in addition to these questions, visits to green markets could be organised to talk to sellers and customers about the practices that these two groups develop in this context. How do the sellers calculate the amount to charge the customers? Are these methods the same as those that students use at school? By looking into these questions in depth, both John and the apples from the semi-reality exercise could lead to the discussion of cultural, social and economic issues by the teachers and students in mathematics classrooms.

To exemplify an exercise with reference to a reality, I presented the following situation to the prospective teachers: "The absence of 30% in the Exame Nacional do Ensino Médio (ENEM)[1] in 2016 was 2.4 higher than 2015, but it followed the average of the historical sequence of the exam. In 2014, the absence was 28.9%; in 2013, 29.7%; in 2012, 27.9%. In 2011, 26.4% of candidates failed to take the exam. What was the absence in 2015? Construct a column chart to represent these data". With this exercise, the teacher wants the students to use mathematical concepts and techniques, such as rule of three, percentage and proportion, to calculate the absence rate in the ENEM in 2015. In addition, they need to construct a column chart in order to realise that the absence rate of the exam was consistent in the last six years. When this exercise is developed in high school classes, certain comments and questions might be made by the students: "Hmm, I do not know what course I want to do! I do not even know if I am going to university. I feel pressure from society and my parents to go to university. Why does everyone have to go to university? In 2016, some schools were occupied. A friend of mine

---

1   The Exame Nacional do Ensino Médio (ENEM) is a tool of evaluation—passing the exam can allow the student to be admitted to certain Brazilian universities.

could not do the ENEM because his school was occupied! Why were they occupied?"

If the teacher considers some of these comments, the students could enter into landscapes of investigation on different themes. One possibility is to discuss the need and pressure to take a higher-level course. Are there vacancies in Brazilian universities for all adolescents? What other possibilities of jobs exist that do not require higher education? Which careers are most promising? What does a promising career mean? Another possibility is to reflect on the protests that occurred in Brazil against the approval of the Proposed Amendment to the Constitution (PEC) 241 or 55 in the second semester of 2016. Why were some schools occupied, causing some students to be unable to take the ENEM? How does the content of that PEC affect Brazilian education? What do the rates and numerical information mean? With these questions, teachers and students can use mathematics to understand what is behind those laws and enter into political, social and ethical discussions.

Until this moment in the course, the prospective teachers had heard and discussed some issues of CME, such as examples of investigative activities and qualities of critical learning. A first step had been taken in the direction of the prospective teachers becoming aware of the existence of CME. However, these actions were not enough to encourage them to think about implementing investigative activities in their future teaching practice. Then, a second step was taken. "Now, it is your turn to open an exercise!" I told them.

In groups, the prospective teachers had to choose an exercise from a textbook with reference to pure mathematics, semi-reality or reality. In a second step, they should open this exercise in order to transform it into an investigative activity, describing the process of transformation and indicating what they had considered when creating that activity. Finally, they should write down the investigative activity they had created.

This task would be an opportunity for the prospective teachers in the course to imagine themselves as teachers in a moment of lesson planning. They would be experiencing the practice of turning an exercise into an investigative activity. In the next section, some reflections on prospective teachers' responses to the requested task will be made.

3. About the Movement from the Paradigm of Exercise to the Landscapes of Investigation

On the task of transforming an exercise into an investigative activity, fourteen responses were given by the prospective teachers. I analysed these answers with some questions in mind: What were the exercise references chosen by the prospective teachers? In the process of transformation, what were the aspects they considered? What were the references chosen for the investigative activity? What were the characteristics the prospective teachers attributed to an investigative activity? The answers to these questions were organised in a table and the important sentences of each task were highlighted.

The exercises chosen by the prospective teachers were taken from textbooks that are used in Brazilian schools. Of the total number of exercises, five (36%) referred to pure mathematics, seven (50%) to a semi-reality and two (14%) to reality. One hypothesis to be considered as an explanation for the majority of prospective teachers choosing exercises with reference to a semi-reality is the great presence of the contextualisation of mathematical concepts in the textbooks, following the government's recommendations for the mathematics curriculum in schools.

Some groups of prospective teachers presented justifications for their reference choices according to Skovsmose's text and the discussions had in the class. An example of this is what Group A wrote: "The exercise [chosen] *just* asks for the calculation of the arithmetic mean of the ages and heights of a sample of 12 players" (Group A, 2018, my emphasis). The adverb "just" indicates that there is only a direct and immediate calculation to be made by the students, characterising the exercise as an application of a technique. Also in this sense, another group pointed out that the exercise chosen had only one possible answer; "all the relevant data for resolution were present in the exercise and the rule to convert litre to millilitre was presented to the student throughout the pre-exercise text" (Group B, 2018).

Regarding the process of opening an exercise—that is, transforming it into an investigative activity—it is possible to note some common characteristics among the aspects pointed out by the prospective teachers in this process. First, I will consider that the transformation of the exercise into an investigative activity was, in fact, accomplished in order to characterise aspects of this process. Later, I will discuss the understanding that the prospective teachers had of an investigative activity, looking at the outcome of the transformation.

Creating a landscape of investigation may have to do with showing the students how mathematics is related to other areas of knowledge, and with discovering new mathematical concepts when addressing real facts. One group transformed an exercise about population of bacteria (a reference to semi-reality) into an activity in which students had to discuss how medicines work in the human body. The chosen exercise was:

> An isolated bacterial colony for culture reproduces so quickly that it doubles in volume in the vats every minute. Knowing that in 6 minutes a bowl is completely full, determine in how many minutes the bacteria occupy half of the vat.
> A) 3 minutes
> B) 4 minutes
> C) 5 minutes
> D) 2 minutes

The prospective teachers presented some questions that could be raised by the teacher and students when they work on the exercise:

> What kind of bacteria are these? Do all bacteria reproduce in the same way? And if the number of bacteria doubles every minute, what happens when they take up the entire vat? Do they stop reproducing? Why were these bacteria being cultivated? (Group D, 2018).

According to this group, a landscape of investigation was created to provide an environment for students to "discover new mathematical concepts in contexts related to our reality, involving a relationship with Biology and Chemistry" (Group D, 2018). The goal was to "show how mathematics relates to other sciences and how it helps in the calculus in those areas" (Group D, 2018).

In order to achieve this goal, Group D designed an investigative activity in which the classroom would be split into groups to research and discuss questions like the following:

> Why do we usually take antibiotics every 8 hours? Does it have anything to do with how bacteria reproduce? How does one estimate this period of hours? And what about antibiotics that have to be taken every 6

hours? Do these bacteria reproduce more quickly? How is the estimate that determines how often we take the next dose of a vaccine calculated? How do we calculate our immune system's response to a disease like yellow fever? Is there any relationship between the reproduction of bacteria and the time we take the medicine? How is the dosage of the medicine calculated? (Group D, 2018).

Discussing these questions, the students could use mathematics to understand how medicines are regulated and the reasons for taking them at a certain period of time. This proposed extension from traditional textbook exercises can encourage teachers and students to discuss what is behind a given—and, most of the time, accepted—exercise in mathematics classrooms.

In the same direction, another group of prospective teachers turned an exercise that referred to a semi-reality into an investigative activity, relating the theme of the exercise to the reality of the students: "To turn this exercise into an investigative activity, the students must experiment in their own reality" (Group C, 2018). The group also stated that:

> For the activity to be investigative, the students must investigate and experiment, in order to be able to discuss and reflect on what happens. Then, the activity will not be a simple and mechanical mathematical exercise (Group C, 2018).

It is possible to see that this group knows of some differences between an investigative activity and an exercise: on the one hand, the discussion and reflection on reality and, on the other hand, the mechanical activity, respectively.

Many groups stressed the importance of questioning during an investigative activity. The questions may arise in the development of the activity, indicating new directions to be followed by the students. The questions which announce the investigative activity created by the prospective teachers show an opening for different possibilities of the students' answers. For example, "What can we say about the possible ramp sizes? And about the values of the internal angles formed by the sides of the ramp?" (Group E, 2018, on urban accessibility). Another type of question created for the investigative activities was: "What if this quantity [of coins] is even? What if all the coins were 10 cents?" (Group F, 2018), indicating new challenges in the investigative activity. The questions that begin with "What can we say ...?" and "What if

...?", created by the prospective teachers, are related in some way to the questions in the examples of landscapes of investigation presented in the course when Skovsmose (2000) was discussed.

The author of the questions was an aspect highlighted by the prospective teachers. The exercise brought for discussion by Group C was about a basketball court. The questions created by the author of the textbook were related to the measures presented in a figure of such a court. In the investigative activity created by this group, these questions should be proposed by the students themselves, and not by the teacher or the author of the textbook: "These questions [in the paradigm of exercise] will not be explained by the teacher" (Group C, 2018). The answers to these questions should be used to solve broader questions of the investigative activity: "The students have to think and come to conclusions about what actions they should take to solve the task" (Group C, 2018, on estimating the cost of repainting the basketball court).

Regarding the references of the activities created by the prospective teachers, the distribution was as follows: three (21%) activities with reference to pure mathematics, four (29%) referring to a semi-reality, four (21%) with reference to reality, and three (21%) activities that combined references to reality and semi-reality. The number of activities with reference to pure mathematics decreased when the exercise became an activity considered investigative by the prospective teachers. This is due to the fact that some groups believe that the transformation process should result in some contextualisation. The same happened with the semi-reality exercises. The reference to real life figured in the investigative activities.

From the five exercises with reference to pure mathematics, three maintained the reference after the transformation to an investigative activity. The concern of these groups was that the students could make new mathematical discoveries, discuss issues among their colleagues, create hypotheses to later come to conclusions, make generalisations and apply what was discovered in particular cases.

With the transformation from an exercise to an activity related to landscapes of investigations, a new type of reference emerged. An understanding of this was experienced by one of the groups: "The proposal is to transform it [the exercise with reference to a semi-reality] into a landscape of investigation mixing references to semi-reality and reality (milieus of learning (4) and (6))" (Group B, 2018). The group started from the following exercise:

> I invited 35 people to my birthday party. I calculated that each person would consume two 290 ml bottles of soda. When I went to the supermarket to buy the soft drinks, only 1.5 litre bottles were on sale. How many bottles should I buy? (Do not forget to include me).

The exercise involves subjects as conversion of measure units, multiplication and division. Group B suggested that the teacher raised some questions for students to encourage them to think about other possible conversions: How many cans of soda can fit in a bottle? What is the relationship between cubic centimeter and litre? Why is the volume of soda not measured in cubic centimeters?

Considering another possible direction for the investigative activity, Group B created a situation of selling some products from imaginary supermarkets, characterising a semi-reality, and they posed questions about whether a certain purchase of different sodas was worthwhile to make if the products had volumes measured in different units. The group also suggested that the students do price research in the supermarkets in the city where they lived, indicating reference to reality. With these activities created by the prospective teachers, the students could use mathematics to make decisions and create new conversions, instead of assuming the truth from traditional mathematics classes. As the group said, the created investigative activity was a mixture of references.

Another understanding for the new reference was brought by two other groups. The activities created began as follows: "The school needs to repaint the basketball court" (Group C, 2018) and "One school comes to know that in the next year some students with reduced mobility will join the school" (Group E, 2018). These schools, at the time the investigative activities were created, did not exist. They were an assumption, characteristic of the semi-reality. At the same time, the appeal for reality was very strong in these two activities. Group C stated that the students "must experiment in their own reality" and, therefore, decided that the best thing for the activity they created was "to measure the basketball court that is in the school area" (Group C, 2018). The prospective teachers considered the existence of a basketball court in the school where they would set the activity. Group E did not state explicitly the reference to reality, but brought images of ramps in real urban spaces and discussed urban mobility.

This type of reference was due to the very characteristic of imagination that the task had. The prospective teachers had to imagine how they would make the transformation. Thus, the teacher, the students, the school and the lessons were created in the context of imagination. The reference that emerged, therefore, can be considered as an imagined reality.

The adjective "investigative" was clarified in different ways when the activity was presented. Sometimes the adjective was made explicit by the authors, sometimes not. It seems that the prospective teachers knew that to be investigative, the activity could not only ask the students to do direct calculations, which is a strong feature of the paradigm of exercise. Some calculations may appear in an investigative activity, but the activity should involve the students in other actions too.

Then what would an investigative activity be to the prospective teachers of the course? Analysing their answers, the adjective "investigative" may be related to creating empirical models to represent a situation, observing and taking notes about a situation, interviewing people, comparing and interpreting information, reflecting on the meaning of a concept, discussing and presenting ideas, making discoveries, raising hypotheses, making decisions, posing questions and contextualising. It seems that, for the prospective teachers, there are notable differences between tasks in the two columns of Skovsmose's matrix.

4. Final Remarks

The discussion of Skovsmose's paper and the task of transforming an exercise into an investigative activity represented a first step in the prospective teachers' education in the context of CME. The aim was to reflect with them on how to transform exercises from school mathematics tradition into investigative activities with different references, in order to develop mathemacy and promote mathematical discoveries by the students in the schools. One can read the examples of transformations designed by the prospective teachers and not see much of political, social, ethical and cultural issues proposed into them. That is why the transformation task represented a first and possible step towards CME by the prospective teachers.

This process of transformation was based on what Skovsmose (2011) calls opening an exercise. In order to open an exercise with reference to pure mathematics, one can explore the theme proposed in the exercise in different directions, following Skovsmose's orientation. In order to open an exercise with reference to semi-reality and reality, I have proposed the exploration of the comments produced by students as they work through the exercise. Through one procedure or another, the exercise changes. The result of the transformation process carried out by the prospective teachers in the course indicated that the investigative activity should have a strong appeal to reality and semi-reality, besides mobilising other actions that are not developed when one solves an exercise, such as interpreting information, discussing and presenting ideas, making discoveries, reflecting and making decisions, and posing questions. Due to the characteristic of the task, a new reference emerged: an imagined reality.

The activity developed was a first moment in the education of the prospective teachers in the context of the CME. This was a possible way for the prospective teachers to approach this theme. A next possible step could be to implement the created activities in the classrooms at schools. This is what I intend to do in another opportunity to contribute to the reflection on how the prospective teachers enter into CME.

Acknowledgements

This chapter, with the exception of the examples of exercises and related investigative activities, as well as its discussion—created by the prospective teachers—was published in the proceedings of the Tenth International Mathematics Education and Society Conference, in 2019, in India (Milani, 2019). The editor of the proceedings has authorised the republication of the text with these additions. A short version of this chapter was published in the journal *Perspectivas da Educação Matemática* in Portuguese (Milani, 2020). The journal's editors have authorised the publication of this expanded version of the original text.

References

Alrø, H. & Skovsmose, O. (2004). *Dialogue and learning in mathematics education: Intention, reflection, critique.* Kluwer Academic Publishers.

Marcone, R. & Milani, M. (2020). Educação matemática crítica: um diálogo entre sua gênese nos anos 1970 e suas discussões em 2017 no Brasil. *Revista Paranaense de Educação Matemática, 9*(20), 261–278. https://doi.org/10.33871/22385800.2020.9.20.261-278

Milani, R. (2019). *Opening an exercise: mathematics prospective teachers entering in landscapes of investigation.* 10th International Mathematics Education and Society Conference. Hyderabad, India. https://www.mescommunity.info/proceedings/MES10.pdf

Milani, R. (2020). Transformar exercícios em cenários para investigação: uma possibilidade de inserção na educação matemática crítica. *Perspectivas da Educação Matemática, 13*(31), 1–18. https://doi.org/10.46312/pem.v13i31.9863

Skovsmose, O. (2000). Cenários para investigação. *Bolema, 13*(14), 66–91.

Skovsmose, O. (2001). Landscapes of investigation. *ZDM—Mathematics Education, 33*(4), 123–132.

Skovsmose, O. (2007). Mathematical literacy and globalisation. In: B. Atweh et al. (Eds), *Internalisation and globalisation in mathematics and science education* (pp. 3–18). Springer.

Skovsmose, O. (2011). *An invitation to critical mathematics education.* Sense Publishers.

17. The Impact of Income Tax on the Teaching Profession:
A Debate Involving Social Justice

*Renato Douglas Gomes Lorenzetto Ribeiro,
Daniela Alves Soares, Adriana de Souza Lima,
Lucicleide Bezerra and Edyenis Frango*

We present in this chapter a theoretical study and an example of teacher education practices, both inspired by the landscapes of investigation and critical mathematics education of Ole Skovsmose, as well as the mathematics for social justice of Eric Gutstein. This approach to teacher education was used with a group of undergraduate students in mathematics and Master's students from a public university in the southeast of Brazil. The chosen theme was related to the teachers' own issues and struggles. We explored a bill by the Brazilian Senate (445/2012), which proposes that teachers of all levels of education be exempt from paying income tax. With this programme, we intended the participants to experience landscapes of investigation in an open space for learning mathematics through a theme involving social justice, and to be able to consider the importance of this kind of approach for their own professional practice in the classroom. Based on the data produced, this chapter presents two narratives as a guiding thread for the analysis of the work carried out, showing that the assumptions of the landscapes of investigation and social justice served as a background for the teacher education proposal.

Teacher education, pre-service and in-service, often presents teachers with the possible pedagogical paths to teach mathematics, but does not always create spaces that allow teachers to experience these paths for themselves as learners. Starting from a teacher education course promoted and offered by *Grupo 5*,[1] we present part of the results of our research which sought to create spaces in which the teachers themselves could experience *landscapes of investigation* (Skovsmose, 2011) from the point of view of *teaching mathematics for social justice* (Gutstein, 2007, 2012).[2]

The course had the support of a public university located in southeastern Brazil, which provided the space for us to carry out teacher education on its premises. This event was open to all interested undergraduate and Master's students in the field of mathematics. The date of the event was set according to our availability to travel to the location, since the members of *Grupo 5* live in four different states: Minas Gerais, Pernambuco, São Paulo and Rio de Janeiro.

The period in which we began to prepare this project for the teacher education course, in April 2018, coincided with the deadline for Brazilian taxpayers to submit their annual income tax report. At the same time, there was a vivid debate in virtual groups about Senate Bill 445/2012 (Brasil, 2012), proposing income tax exemption for teachers active in all levels of education, which had the support of many professionals in the field. We believed that this topic would be known to the participants, and that they would likely already have an opinion on it, because discussions on social networks were leading teachers to manifest their opinions on the proposal on a page of the Brazilian Senate website.[3]

To understand the interest of a part of the Brazilian population in defending the proposal, it is necessary to consider the context. A Brazilian teacher earns, on average, less than other college-graduate professionals. According to data from the Brazilian Economy Ministry, in 2018 the average monthly remuneration of professionals with a

---

1   The authors of this chapter are members of Grupo 5, a research group formed at the Second Research Colloquium on critical mathematics education, held in April 2018, at the Universidade Estadual Paulista (Unesp) in the city of Rio Claro, state of São Paulo, Brazil.
2   Some of these activities and reflections have already been presented at events in which *Grupo 5* participated (Soares, Lima and Frango, 2018; Frango et al., 2019).
3   https://www12.senado.leg.br/ecidadania/visualizacaomateria?id=109603

college degree was R$6,155.31[4] (Brasil, 2018). In that year, as well as in the previous one, the official national monthly minimum wage for a teacher was R$2,455.35 for a 40-hour working week (Brasil, 2017). It is important to highlight that there are wide salary differences, depending on the level of education in which Brazilian teachers work, and whether it is a private or public school—if it is a public school, whether city, state or federal. Even so, when compared to the average minimum wage of college-graduate professionals in general, the minimum wage of teachers can be considered inadequate.

We considered the controversy surrounding the issue to be a good theme for creating a guiding question able to involve the participating teachers in a debate on the promotion of social justice using mathematical concepts. In fact, working with tax policy in mathematics education is not unprecedented. The theme can be found in the work of several authors; one example is Marilyn Frankenstein (2012), who had expressed her intention to work systematically with this theme in low-income communities. Tax policy seemed to her to be an appropriate theme given her goal of engaging people in the struggle for social justice.

Grupo 5 also saw in this theme a good opportunity for developing a course addressing social justice. In addition to the objective of promoting the experience of a landscape of investigation, our aim in this teacher education project was to create a space for mathematics learning using a theme involving social justice that would be of relevance to the participants. Furthermore, it was our intention for the participants to evaluate the pedagogical purpose of the landscapes of investigation on this subject theme, and to reflect on the possibilities of future application in their classrooms.

In this chapter we present the theoretical assumptions for the course, some of the activities, and an analysis of some critical events that took place during the course.

Finally, it is worth highlighting that some of these activities and reflections have already been presented at events in which *Grupo 5* participated (Soares, et al., 2018; Frango, Lima and Soares, 2019). Below,

---

4   Equivalent to US$1,098.79 and €1,003.10. The exchange rates used for the Brazilian Real to the US dollar and to the euro are: R$1.00 = US$0.1785 = € 0.1630. (www.bcb.gov.br).

we present the theoretical assumptions that are directly connected with the objectives and with our analysis.

1. Landscapes of Investigation

Critical mathematics education (CME) is characterised as a set of concerns regarding mathematics education (Skovsmose, 2008, 2011) that takes into account social aspects of education and proposes, among other things, working with landscapes of investigation for the study of mathematical concepts in the classroom.

According to Ole Skovsmose (2011), landscapes of investigation constitute a new paradigm for classroom practice, as opposed to teaching through mechanical and exhaustive lists of exercises, based on the *exercise paradigm*. Such lists provide the educator with a safe and predictable path, while a landscape of investigation moves teachers—and students—out of their *comfort zones* into a *risk zone*, in which unforeseen situations may arise.

Landscapes of investigation can support investigative work and are characterised as a terrain on which teaching-learning activities take place, favouring the emergence of new possibilities, since the students are the protagonists of the process. However, simply conducting an investigation in the classroom would not characterise a landscape of investigation; there is also the need for each student to accept the invitation to participate in the activity.

CME invites us to take part voluntarily, with no presumption of a formal or concrete invitation, but rather a call to involvement with the theme proposed. Even so, our course also had a physical invitation, a folder with information and the invitation to participate in the landscape called *What Is the Impact of Income Tax on the Teaching Profession?* As we understand it, the people who attended the course are those who accepted the invitation to be involved with the theme advertised.

In the introductory chapter of his book, Skovsmose presents a diagram with types of references which a mathematics teacher can use in classroom activities (references to pure mathematics, semi-reality and real-life). All these references can be in the exercise paradigm and in the landscapes of investigation resulting in six environments: (1), (2), ... (6). He points out that he does not expect teachers always to work with landscapes of investigation with references to reality (6)—although this would be the most propitious environment for critical learning—but teachers and students together should seek ways that will take them through the different learning environments, according to the needs of the class.

2. Teaching Mathematics for Social Justice

Teaching mathematics for social justice is strongly related to CME. One of the important authors concerned with issues involving social justice in mathematics education is Eric Gutstein. He argues that teachers need to get close to school communities and, along with them, identify themes involving social injustice that are part of their struggles (Gutstein, 2012); in other words, addressing issues that express cases of *social oppression* (Freire, 1987). Gutstein has a Freirean view of education, which we share. In particular, it seems appropriate to us when Paulo Freire says that reading the world precedes reading the word (2017), because what good would it be to know how to read, but not be able to *read the world*? In the same way that Freire speaks of reading the world, there is also the possibility of *writing the world*. In an interview with Amanda Moura and Ana Carolina Faustino, Gutstein points out that reading and writing the world with mathematics "means essentially that students must use and learn mathematics to study their social reality, so that they can have a deeper understanding of the world and be prepared to change it, as they see fit" (Moura and Faustino, 2017).

Among the main concerns of teaching mathematics for social justice, we can especially highlight the importance of students' socio-political engagement. But how can a mathematics teacher create this socio-political engagement among the students? Gutstein states that "one of the principal ways for teachers to support students in moving toward these interconnected goals is for the students to engage in mathematical investigations in the classroom of specific aspects of their social and physical world" (Gutstein, 2007, p. 109).

When adopting teaching goals that are not limited to academic mathematics, it becomes necessary to reflect more deeply on the knowledge that becomes involved. Gutstein sees three distinct types of knowledge: community, critical and classical knowledge. In his understanding, community knowledge is linked to cultural, local issues—it is usually the knowledge students learn outside the school, individually and with the community, the knowledge they bring to the classroom. Classical knowledge refers to academic, formal and abstract knowledge traditionally taught in schools. Critical knowledge refers to the "socio-political conditions of one's immediate and broader existence. It includes knowledge about why things are the ways that they are and about the historical, economical, political, and cultural roots of various social phenomena" (Gutstein, 2007, p. 110). These three types of knowledge that Gutstein distinguishes between are interconnected when the intention is to develop classes within a perspective of mathematics for social justice, in which teachers open space for discussions on themes that are emancipatory for the student community, in order to enable socio-political engagement, using mathematical tools to enable understanding and action in this environment.

The circumstances and concepts that we have set out so far have guided us to choose the theme, and outline the activities, which in our view would transit through the various learning environments and make use of the different types of knowledge (classical, community and critical). Next, we present part of the collaborative work carried out, which has proved to be a fertile path for experiences and discoveries from the point of view of CME and teaching mathematics for social justice.

3. What Is the Impact of Income Tax on the Teaching Profession?

As is true of other Latin American countries, Brazil has profound inequality of access to public services. According to Alberto Carlos Almeida (2010), many Brazilians believe that corruption is responsible for the poor provision of public services to the population and that a reduction in taxes would generate more jobs, increase consumption, favour economic development and increase tax collection. In Almeida's opinion, these beliefs contribute to Brazilians' aversion to paying taxes.

According to Eduardo Fagnani and Pedro de Carvalho Jr., the Brazilian tax system is responsible for promoting a context that penalises the middle- and low-income groups, reduces families' disposable income, reproduces inequality, weakens aggregate demand and, consequently, limits economic growth and the country's development (2019, p. 38). These authors also point out that the effective tax burden in Brazil, which was 32.4% of gross domestic product (GDP) in 2015, is not high when compared internationally. Indeed, it is lower than the average for the member countries of the OECD (Organisation for Economic Co-operation and Development): 34.1% of GDP. However, on average, the proportion of taxes on consumption is much higher in Brazil: 49.7%, in comparison with 32.4% in those countries.

Such observations show the need for a debate about the Brazilian tax model, including the proposed income tax exemption for teachers as a way to compensate for their low wages. It raises the question of whether, in fact, such an exemption would be a form of social justice and would contribute to the adoption of tax measures beneficial to the country. It is worth highlighting that the participants in *Grupo 5* already had their own opinions on the topic. All were against passing the law, although they did not express this opinion during the course.

Before going on to describe our course, we briefly present the main guidelines that regulate the calculation of personal income tax in Brazil.

4. Income Tax in Brazil

Although the Brazilian tax authority (the *Secretaria da Receita Federal* or Federal Revenue Service) makes a computer app available for taxpayers to enter their data and obtain the revenue due, for ordinary people the calculation of income tax due is quite complex. There are different rates for different ranges of taxable amounts. The rates shown in Table 1, levied on the income of all professional activities, are not applied on the taxpayer's gross or net income, but on a "taxable amount" (referred to as the "calculation basis").[5] This amount comprises the taxpayer's actual income, minus a series of deductions all specified by law:

CALCULATION BASIS = REMUNERATION[6] – LEGAL DEDUCTIONS

---

5 It is defined in Article 44 of the national tax code, as the amount—real, arbitrated, or presumed—of taxable income or earnings (Brasil, 1966).

6 In Brazil, the term "remuneration" is the amount paid to a worker by a private company or by the government, in case of a government employee.

Table 1. Progressive Table for Calculation of the Monthly Personal Income Tax to Be Paid. Reproduced from Brasil (2017).

Monthly Calculation Basis (R$)	Amount to Deduct (R$)	Rate (%)
Up to 1,903.98	-	-
From 1,903.99 to 2,826.65	142.80	7.5
From 2,826.66 to 3,751.05	354.80	15.0
From 3,751.06 to 4,664.68	636.13	22.5
Above 4,664.68	869.36	27.5

The deductions are amounts which taxpayers are entitled to deduct from their remuneration to arrive at the taxable amount. Examples of the most frequent deductions are: deductions for declared dependants, healthcare and education expenses, social security contributions, social benefits received, alimony payments, etc.

For example, for a person who, after deductions, has a taxable amount ("calculation basis") of R$ 4,000.00 (four thousand reais), we can calculate the tax due, as shown in Table 2:

Table 2. Income Tax Calculation for a Taxable Amount of R$4,000.00.

Tax Rate	Calculation	
	Amount in this range (R$)	Tax due in this range (R$)
Exempt	Up to 1903.98	Exempt = 0.00
7.5% range	2 826.65 – 1 903.98 = 922.67	7.5% of 922.67 = 69.20
15% range	3 751.05 – 2 826.65 = 924.40	15% of 924.40 = 138.66
22.5% range	4 000.00 – 3 751.05 = 248.95	22.5% of 248.95 = 56.01
TOTAL	R$4000.00	R$263.87

The tax authority also presents a direct way to calculate the tax due, as shown in Table 1, introducing the amount of the instalment to be deducted, as a way of simplifying the calculation. In our example, instead of calculating tax due for each range in the table, we substitute this process by calculating the tax due only for the range in which the calculation basis is found, subtracting the amount to be deducted. For the amount of R $4,000.00, the tax calculation would be performed as follows:

$$\underbrace{4000}_{calculation\ basis} \times \underbrace{0.225}_{rate} - \underbrace{636.13}_{amount\ to\ deduct} = \underbrace{263.87}_{tax\ to\ be\ paid}$$

5. The Teacher Education Course

The proposed course was planned to last eight hours, spread over two days, and was offered at a Brazilian federal university in south-eastern Brazil. The invited participants were undergraduate and graduate students in mathematics at that university.

At the time of the course, the proposal to exempt remuneration of teaching activities from income tax was still being considered in the Brazilian Senate. This proposal guided our teacher education course based on the following question: "Are you in favour of income tax exemption for teachers? Why?"

The participants formed groups to carry out five activities over the two days. The activities were as follows: the first aimed at discussing what income tax is, and how it is calculated in Brazil; the second activity was a presentation of a teacher's payslip, with discussion of the calculation of both the income tax and the social security contribution;[7] the third aimed to challenge a common occurrence in the teaching profession which interferes with the amount of taxes paid (having two or more sources of income); the fourth activity aimed to present the various bills that had dealt with an income tax exemption for teachers; and in the fifth and final activity, participants were able to identify and calculate possible financial effects on the public purse in Brazil if the income tax exemption for teachers was to be approved. Two of these activities will be detailed further below in terms of first and second narrative.

---

7 We addressed the subject because the participants did not know how the social security contribution was calculated.

6. Data Presentation and Analysis

We collected the data using a questionnaire, photographs, and video and audio recordings of the course. The research was qualitative in nature and our choice of method, as suggested by Arthur Powell and Wellerson da Silva (2015), was to consider video records as being the main source of data, rather than the transcription of audio.

Then, we transcribed the dialogues and built narratives, preserving some original dialogues of moments when the theoretical aspects studied could be observed. We called these moments *critical events*, as we understand that, in them, we notice a change in the understanding of students, showing a conceptual advance related to the hypothesis of the investigation (Powell, Francisco and Maher, 2004).

After careful analysis of the videos, individually and collectively, we identified the critical events, which culminated in the narratives that follow. We selected two critical events in which the discussions contributed strongly to the development of the landscape of investigation. In the following we concentrate on these two events.

The course ended with the same question as it started: "Are you in favour of income tax exemption for teachers? Why?" We asked the participants to use the knowledge we focussed on during the course to expand the justifications to their answers to this question.

It is important to highlight that some questions were more open and reflective ("Is the income tax table fair?"), while other questions were more closed ("Calculate the amount to be deducted for a calculation basis of R$3000,00"); some of the activities involved references to reality ("A real teacher's payslip"), and other activities involved pure mathematics ("Determine the function that would calculate the tax"). The different approaches and directions given to each proposed activity were intended to make participants transit between the different learning environments, as recommended by Skovsmose (2011).

6.1 First Narrative

On the first day of the course, three participants were present. After an initial conversation in which they and the course trainers (Adriana Lima, Daniela Soares, Edyenis Frango and Renato Ribeiro) introduced themselves and presented the goals and expectations, we showed a video (Schimidt, 2011) in which the origin and some forms of taxes in

Brazil were discussed. This video also contained a report on the estimate of taxes embedded in the price of food, paid by a woman who had a monthly income of 300 reais. We then made a comparison of the amount of tax paid by this woman and her income, concluding that, in Brazil, the rich end up paying a lower proportion of their income in taxes than the poor.

After showing the video, we presented and briefly explained the income tax table (Table 1), and then proposed Activity 1, asking participants to calculate the amount of monthly tax due for the following taxable amounts (calculation bases): R$1,500, R$2,500, R$3,500 and R$4,500.[8]

The participants solved the calculations, discussed among themselves the strategies adopted, and then explained their answers and how they reached them. During the discussion, one participant asked why calculation of income tax is not a subject covered in the university discipline of financial mathematics, saying: "Don't you think it is important to learn this?" (Fig. 1).

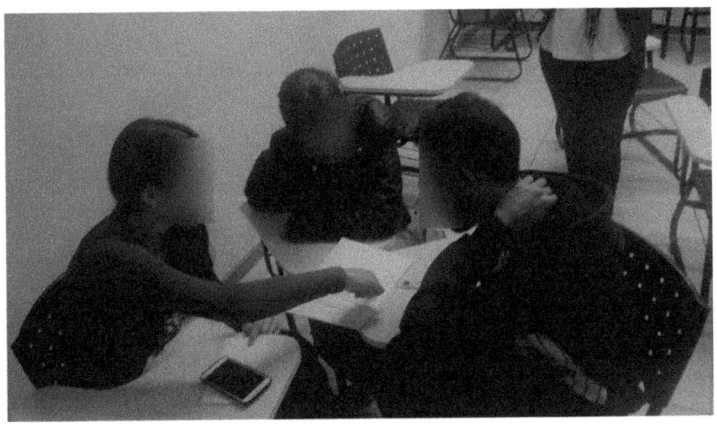

Fig.1. Participants performing the activity.

In the second part of Activity 1, we asked the participants to calculate the amount to be deducted. When presenting the exercise, the trainer asked the participants where the number came from. The plan was to

---

8   The amount of tax due for each taxable amount (basis of calculation) is, respectively: R$0.00; R$44.70; R$170.20; R$376.37. It is based on these amounts that the participants evaluated Table 1 during the dialogue transcribed at the end of the first narrative.

allow time for the participants to solve the question together, and then present the result, and the strategy adopted, on the blackboard (Fig. 2).

Fig. 2. A participant explaining how their group arrived at the result.

Then, one of the trainers (Renato) asked the participants if the tax table in question was fair. Participants 1 and 2 and trainer Daniela made comments (Participant 3 did not respond). Then, trainer Renato said that they should discuss the issue between them and present a verdict together. After some time, they came up with an answer. The following is a transcript of the dialogue:

> *Renato*: Have you reached a verdict yet? Yes or no?
> *All*: Yes.
> *Renato*: OK! Then you can explain it to us.
> *Participant 1*: We came to the conclusion that it is not fair.
> *Renato*: It's not fair?
> *Participant 1*: Exactly.
> *Participant 2*: Also, we don't really know what a fairer way would be...
> *Participant 1*: Because it is not proportional ... while the salary increases, the tax does not grow in proportion to the increase in salary.
> *Renato*: Um ...
> *Participant 1*: Yes, it's basically that ...
> *Participant 2*: It practically doubles ...
> *Participant 1*: Yes, the percentage (tax as a percentage of salary) practically doubles ... in the first salary (this percentage) is 1%, right ... 1.7% ...
> *Participant 2*: Yes ... one point something ...
> *Renato*: Yes ... 2% would be 50, right?

Participant 1: And ... at 3,500 it goes to four point something, and then 8% ... and at 4,500 it's four point something ... the tax ... the percentage of the salary.
Renato: Almost 5%, right, because 10% is 350, half, 175 ...
Daniela: And why isn't it fair?
Participant 2: Because if you earn an extra 1,000 reais... but the amount of tax has doubled ... not even 1,000, 900.
Participant 1: If the salary keeps increasing, there will come a time when it will match the number ... I think.
Renato: Huh ...
Participant 1: From what we're seeing there, it's growing so much ... if, for example, it continues in the same pattern ...
Participant 2: If you continued the table, if it didn't stop, the tax grows much more than your salary.
(*Group 5*, Research Data).

The dialogue on this subject continued for a while, then the group started the next activity.

6.2 Analysis of the First Narrative

The initial discussions were open and allowed the participants to freely seek ways to reach the solution. The more closed activities were complementary or intended to support some discussion of the open questions. In this way, we were able to identify that the activity was associated with different learning environments. The first part of Activity 1, in which one learns to calculate the tax due, referred to real life, approximating to Environment (5). The second part, in which we asked participants to calculate the amount to be deducted, made reference to pure mathematics (1); although they used real data, to answer the question it was necessary to determine the rule(s) underlying the function, and this brought up the question of why that number could facilitate manual calculation.

As to how the participants related to the activities, we identified situations we consider to be closely associated with the teaching of mathematics for social justice: the use of mathematical concepts to justify an argument related to a social issue, the use of non-mathematical arguments, and allocation of value to critical and community knowledge.

With this in mind, we identified some critical events. The first was the moment when the participants explained why they considered the tax table to be unfair. They stated that if a person's taxable amount (calculation basis) increases by 1,000 reais, the tax due will also increase by a certain amount, but if the calculation basis increases by 2,000 reais, the increased tax will be more than double the tax in the first situation (since for a calculation basis of R$2,500.00 the income tax is R$44.70, for R$3500 it is R$170.20, and for R$4500 it is R$376.37), and thus, since the increase of tax is not proportional to the increase in calculation basis, the table is not fair. We did not expect them to make explicit use of a mathematical concept so quickly to justify their stance on a social issue, as our question was too open.

Just as unexpected as this precocious use of mathematical concepts to justify an opinion was the absence of use of non-mathematical arguments. They considered that evidencing the lack of proportionality between the quantities value of tax due and value of the calculation basis would be sufficient to determine whether the table was fair or not. In our view, the participants, in their justification, gave value to what Gutstein (2007) calls classical knowledge, but did not give the same emphasis to their own community knowledge.

At the very start of the course, we watched and discussed a video about taxes, and the participants expressed a clear discomfort when they became aware of the tax burden borne by a person who only had 300 reais per month; still, they did not use this information when they justified their answer about the fairness of the table. Bearing in mind that the current table exempts earnings of up to R$1,903.99 from income tax, if we adopted only the proportionality criterion to create a way to calculate income tax, it would be equivalent to having a single tax rate for all levels of income. The direct consequence of this would be that this person, who has an income of only 300 reais, would be required to pay income tax.

This first critical event raised a new question that we still need to investigate in detail: how can the sequence of activities within a teacher education course influence the establishment of connections between the three types of knowledge: community, critical and classical?

The use of mathematical concepts as a form of justification, at such an early stage of the discussion, led us to consider that the participants may have been influenced by the activity that asked them to calculate the tax values, since they used the same values proposed in the exercise to justify the answer. But why was it not possible to see any influence from the video and the subsequent discussion? We must consider the hypothesis that they, being from within an academic environment and knowing that we were doing research in mathematics education, may have been influenced to express their mathematical knowledge. This is one of the aspects to be considered when designing future courses.

With regard to the appreciation of community and critical knowledge, one of the participants questioned the choice made by the university itself not to incorporate the subject of taxes within financial mathematics. Based on our framework, we interpret that she considers that teacher education should not focus exclusively on classical knowledge, since the knowledge that people already have about their own financial lives would allow for a better analysis and a new perception of reality, using mathematics. This view coincides, therefore, with the notion of reading and writing the world proposed by Freire and Gutstein.

6.3 Second Narrative

On the second day of the course, we received a new participant (Participant 4). Although he did not participate in the first day, he brought knowledge from the experience of helping, in practice, his family members to file their income tax returns. Before starting the activities planned for the day, we shared with the new participant the activities experienced on the previous day. We then discussed some comments posted on social networks about the acceptance (or not) of the bill to exempt teachers from tax. Participants were led to reflect on these comments.

We also showed a history of the various bills that have dealt with the topic,[9] including the one that inspired our course, Senate Bill 445/2012:

Summary: Grants income tax exemption on the remuneration of teachers, under the conditions it establishes.

---

9   The first, dated 2011, was set aside with the justification of it not being appropriate to the budget. The second, from 2012, was still under consideration in the Senate until the beginning of 2019 (Senate Bill 445/2012). The third dates from 2014, and was rejected on the grounds of not guaranteeing tax equality as established by the Constitution; the most recent, from 2018, proposed a change in the constitution, in Item II of Article 150, but was also discontinued after the renewal of the legislature that year.

Explanation of the Summary: Amends Law 7713/1988 [the income tax law], to exempt from income tax the amounts received, as remuneration, when the beneficiary is a teacher currently working in early childhood education, primary, secondary, or higher education. It provides that this law comes into force on the date of its publication and will have financial effects from the first day of the year following that of its publication (Brasil, 2012, our translation).

After this presentation, participants were asked to assess whether this bill would have a real chance of being approved. We oriented them to the texts of Senate Bill 445/2012 carefully, together with Article 150 of the Brazilian Constitution (Brasil, 2016), items I and II:

Article 150: Without prejudice to any other rights of taxpayers, the Union, the states, the Federal District, and the municipalities are forbidden to:
I — impose or increase a tribute without a law to establish it;
II — institute unequal treatment for taxpayers who are in an equivalent situation, it being forbidden to establish any distinction by reason of professional occupation or function performed by them, independently of the juridical designation of their incomes, titles, or rights (Brasil, 2012, our translation).

Initially, the participants presented arguments such as "There is money for that, you only need to improve financial management", and "I support this project and I think it should be approved", among others. After that, when advised to try to justify their comments on the basis of the texts delivered, they expressed the opinion that the Senate bill should be approved, and the following critical event followed:

Participant 4: "Instituting unequal treatment for taxpayers in an equivalent situation" ... In this case, the situation would be related to ... to teachers, in this case. So, if it meant teachers at public education institutions only, it would violate this article [Article 150]. I think that the way it is, it could succeed [in being approved].
Renato: As I understand it, what he said was: if it referred to teachers of public schools only, then it wouldn't be allowed. But since all teachers are included, there is isonomy.
Adriana: Ah, because he read the part where the issue of equivalence is written, right? People in equivalent situations. If you exempt one group, and you do not exempt the other, you will be violating isonomy.
Daniela: Interesting.
Adriana: But then ...
Participant 4: ... Then comes the next part: "any distinction by reason of professional occupation" is forbidden.

Adriana: This exemption is being proposed for what reason? [a few seconds pass ...]
Daniela: It is interesting what he said, because it says: "institute unequal treatment for taxpayers who are in an equivalent situation". As he interpreted here that, among all teachers, you cannot treat them unequally, for example, teachers at public and private educational institutions. So, I am not really reading here "among all the taxpayers in the nation", you know ...
Edyenes: Any category, any ...
Daniela: Exactly, he meant this group of taxpayers [the teachers] and said, "any distinction on the grounds of ... is forbidden", but always having this group in mind.
[further on, other interpretations followed, and the last dialogues of this passage were:]
Daniela: But in the law, there will always be doubts ...
Adriana: There are loopholes!
(*Grupo 5*, Research Data)

After this dialogue, the course continued, with a new activity.

6.4 Analysis of the Second Narrative

A landscape of investigation is also characterised by the presence of risks (Skovsmose, 2011). The teacher may put challenges to the class, having no control over the answers or the content. We understand that the second narrative presents exactly this characteristic.

When we showed the participants the various bills that had addressed the topic of income tax exemption, and the reasons given to justify their having been set aside or rejected, we aimed, albeit implicitly, to present the contradictions found in those bills. When we proposed the reading of Senate Bill 445/2012 and Article 150 of the Constitution, we imagined that the contradiction between the texts would be explicit. It was our view that perhaps at that point the participants would identify impossibilities for application of the proposed law. However, in an open scenario there are many possibilities for answers, and indeed this activity was no exception for two reasons—first, due to the weight of the participants' personal experiences and opinions, and second, due to the laws with their multiple interpretations.

In order for the participants to answer our question about whether the bill should be approved based on its text and whether it was in compliance with the Constitution, they read the material carefully. Even so, when questioned, they answered with personal arguments that were not related to the material delivered. From this fact, we understood that personal experiences and opinions often weigh more than arguments based on the law.

In a second moment, when led to use the material we provided to them, to express their justifications, the participants again surprised us. From reading the excerpt of Article 150 of the Constitution, which says "it is forbidden [...] to institute unequal treatment for taxpayers who are in an equivalent situation, or to establish any distinction by reason of professional occupation or function performed by them" (Brasil, 2016), they had a different interpretation than we had when reading the same excerpt. The concept of isonomy became a key issue. For us, the passage would make it impossible to exempt teachers from income tax because the Constitution does not allow a distinction to be made between people of different occupations (teachers and other professionals). For the participants, as evidenced in the dialogue, their interpretation was that the distinction prohibited was between people in equivalent situations. In addition, they understand equivalents as being part of the same profession—i.e., the exemption would be possible as long as it served all classes of teachers, both from public and private education institutions. This interpretation is one that we did not expect.

7. Final Considerations

We consider that we have made positive progress in our objectives of providing participants with an experience of a landscape of investigation, promoting a space for learning mathematics through a generative theme that involves social justice and leading participants to reflect on the relevance of using landscapes of investigation in their teaching practices.

Our analysis corroborates each of these points, both in that the participants engaged with the proposal, and also in that they voluntarily verbalised dissatisfaction with a teaching practice that covers only classical knowledge.

The proposed landscape of investigation allowed the participants to explore different perspectives, and it directed the course proponents to a risk zone—that is to say, an environment that was largely unpredictable and little-controlled—going through the various learning environments.

However, many discussions, including those highlighted in the above analyses, were concentrated on an approximation of the landscapes of investigation—Environments (2), (4), and (6)—which, in our view, constituted a break from the exercise paradigm.

For critical mathematics education and for teaching mathematics for social justice, any topic that is relevant to the community could be explored. We thus understand that we could have chosen a theme other than "bill proposing income tax exemption for teachers at all levels of education" (Brasil, 2012, our translation). However, the theme was well-accepted in the course that we carried out, and we see great potential in the activities that we set out.

From the landscape of investigation that we prepared, the participants had the opportunity to learn about how to calculate income tax and to understand its function, the impact on the federal budget, and the legal aspects of the proposed exemption bills. However, further to that, they were able to reflect on whether or not it was fair to pay this type of tax, and to think of alternative proposals, using mathematical knowledge.

The course participants responded to a final assessment, with questions about whether the course reached the expected objective, about the pertinence of the practical approach, the consistency of the theoretical approach, the length of the course, the activities proposed, the teaching material, the effectiveness of the learning process, and whether they believed they could apply the knowledge acquired during the course in their professional practice. In general, each of these points was assessed as satisfactory—in most cases—or partially satisfactory. The number of hours was unanimously considered to be insufficient. There is evidence, therefore, that the participants welcomed the experience and identified the importance of this type of work.

It was clear that some aspects must be taken into account for the next course. One of them is a resizing of the total number of hours, as well as adopting a more flexible dynamic that allows an expansion of the debates arising from the reflections that emerge, without the activities being prioritised over the discussions.

Even so, the points perceived during this teacher education course lead us to agree with Gutstein (2012), when he highlights the need for a teacher to have many experiences like this in the process of learning and teaching mathematics for social justice—that is to say, for the improvement of our actions. Consequently, this experience was important both for the implementation of new courses like this, and also in complementing existing research in critical mathematics education and teaching mathematics for social justice.

Clearly, in relation to understanding the Brazilian Constitution, we entered a field in which we did not have much expertise. Yet however much we had studied, we would not have been likely to cover all the possibilities. Perhaps a legal professional would deal differently with the participants' interpretation of the text of the law. At any rate, it was an open landscape not only for the participants, but also for us, the teachers. And that is perhaps the greatest value of this work: allowing oneself to enter areas beyond the strict field of mathematics, beyond closed questions, exploring possibilities, learning together and taking risks.

References

Almeida, A. C. (2010). *O dedo na ferida: menos imposto, mais consumo*. Record.

Brasil. (1966). *Lei n. 5.172, de 25 de outubro de 1966*. Dispõe sobre o Código Tributário Nacional e institui normas gerais de direito tributário aplicáveis à União, Estados e Municípios. Brasília. 1966. Retrieved July 20, 2018, from http://www.planalto.gov.br/ccivil_03/leis/l5172.htm

Brasil. (2012). *Projeto de Lei do Senado PLS 445/2012*. Concede isenção do Imposto de Renda sobre a remuneração de professores, nas condições que estabelece. Retrieved August 02, 2018, from https://www25.senado.leg.br/web/atividade/materias/-/materia/109603

Brasil. (2016). *Constituição da República Federativa do Brasil de 1988*. (*1988, Oct 5*). Brasília: Senado Federal, 2016. Retrieved May 31, 2018, from https://www2.senado.leg.br/bdsf/bitstream/handle/id/518231/CF88_Livro_EC91_2016.pdf

Brasil. (2017). *Receita Federal do Brasil*. IRPF: Imposto sobre a renda das pessoas físicas. Retrieved July 02, 2018, from https://www.gov.br/receitafederal/pt-br/assuntos/orientacao-tributaria/tributos/irpf-imposto-de-renda-pessoa-fisica

Brasil. (2018). *Relação Anual de Informações Sociais—Rais 2018*. Brasília. 2018. Retrieved June 29, 2021, from http://pdet.mte.gov.br/images/rais2018/nacionais/2-apresentacao.pdf

Fagnani, E. & Carvalho Jr., P. H. (2019). *Justiça fiscal é possível na América Latina?* PSI.

Frango, E. R., Lima, A. S., & Soares, D. A. (2019). Imposto de renda: uma proposta de cenário para investigação na formação de professores. *Encontro Nacional de Educação Matemática (ENEM)*. Cuiabá, MT, Brasil.

Frankenstein, M. (2012). Beyond math content and process: Proposals for underlying aspects of social justice education, In A. A. Wager & D. W. Stinson (Eds), *Teaching mathematics for social justice: Conversations with educators* (pp. 49–62). National Council of Teachers of Mathematics.

Freire, P. (1987). *Pedagogia do oprimido (17th edition)*. Paz e Terra.

Freire, P. (2017). *Pedagogia da autonomia: saberes necessários à prática educativa*. Paz e Terra.

Gutstein, E. (2007). Connecting community, critical and classical knowledge in teaching mathematics for social justice. *The Montana Mathematics Enthusiast*, 109–118.

Gutstein, E. (2012). Reflections on teaching and learning mathematics for social justice in urban schools. In: A. A. Wager & D. W. Stinson (Eds), *Teaching mathematics for social justice: Conversations with educators* (pp. 63–78). National Council of Mathematics Teachers.

Moura, A. Q. & Faustino, A. C. (2017). Eric Gutstein e a leitura e escrita do mundo com a matemática. *Revista Paranaense de Educação Matemática (RPEM)*, 6(12), 10–17.

Powell, A. B., Francisco, J. M. & Maher, C. A. (2004). Uma abordagem à análise de dados de vídeo para investigar o desenvolvimento das ideias matemáticas e do raciocínio de estudantes. *Boletim de Educação Matemática (Bolema)*, 17(21), 81–140.

Powell, A. & Silva, W. Q. (2015). O vídeo na pesquisa qualitativa em Educação Matemática: investigando pensamentos matemáticos de alunos. In A. Powell (Ed.), *Métodos de pesquisa em Educação Matemática: usando escrita, vídeo e internet*. Mercado das Letras.

Schimidt, F. (2011). *Tributo: Origem e Destino* [video file]. https://www.youtube.com/watch?v=_HIvSlnXpg4

Skovsmose, O. (2001). Landscape of investigation. *ZDM—Mathematics Education*, 33(4), 123–132.

Skovsmose, O. (2008). *Desafios da reflexão em educação matemática crítica*. Papirus.

Skovsmose, O. (2011). *An invitation to critical mathematics education*. Sense Publishers.

Soares, D. A., Lima, A. S., & Frango, E. R. (2018). *Cenários para investigação e educação matemática para justiça social na formação de professores*. II Simpósio da Formação do Professor de Matemática da Região Sudeste (II ANPMAT). Universidade de São Paulo.

18. Critical Mathematics Education in Action: To Be or Not to Be[1]

Paula Andrea Grawieski Civiero and
Fátima Peres Zago de Oliveira

Society today presents a civilising equation where it is crucial to unveil and guide the imbricated relationship between what is technical and what is human. Therefore, the study of contemporary variables is central to the interpretation of this reality. By considering that critical mathematics education (CME) is the most developed approach to treating such themes in mathematics classes, we present a landscape of investigation developed with high-school students, step by step, based on reflective didactic transposition (RDT) of a Scientific Initiation (SI) project in high school. SI enables the investigation of contemporary themes that, in turn, approach the concerns of CME by fostering questions, autonomy, decision-making and a critical interpretation of reality. This proposal evidenced the approximation of SI concepts with landscapes of investigation, just as it was possible to perceive the urgency of the imbrication between the different milieus of learning and the contemporary variables of this complex civilising equation.

To be or not to be, that is the question: will it be more noble.
In our spirit suffering stones and arrows
With which Fortuna, enraged, targets us,

---

1   Part of this paper was first published in Portuguese and English as Civiero and de Oliveira (2020). We thank the editors of *Acta Scientiae* for their kind permission to draw on this publication.

Or rebel against a sea of provocations
And in struggle to put an end to them? Die ... sleep: no more.
(Hamlet—William Shakespeare)

During the last decades, technoscientific development has been reaching unthinking levels, which, in turn, has boosted a new civilising behaviour.[2] This reveals that people need to consume and possess, rather than be. An organised society is susceptible to the commands disseminated by those who dominate the technoscientific apparatuses that, in turn, are treated as instruments of power, and not as a vehicle for human development.

Thus, there is a civilising equation—a metaphor, utilised by Bazzo (2019)—which could be a tool "to bring together the most different variables that arise at all times in a civilisation that is vulnerable to the most accelerated mutations in its daily behaviour" (p. 21). The tool could be even more significant given the implications that these issues have on society. In other words, these issues give us the urge to "provide reflections and changes in our ways of working knowledge in such serious times of human problems" (p. 20). The civilising equation has both more technical and more human contemporary variables. It is aimed at the overlapping of the variables, so that the result of the equation at least guarantees the principles of human dignity.[3] The current social, economic and political variables are considered essential elements for the analysis and interpretation of reality. Some examples are environmental issues, the immigration process, social inequalities, the hybrid crisis, the atomic bomb, global warming, chemical wars, biological wars, pandemics—such as coronavirus—and so many other variables that compose the civilising equation.

We understand that mathematics is especially relevant in knowledge processing, and operates in the process of globalisation,[4] i.e., it interferes

---

2   Civilising Behaviour—behaviour according to the constitution of the current civilisation and governed by social transformations, as a social construct. This understanding is in line with Norbert Elias, who, in *The Civilizing Process* (1994)—original publication in 1939—analyses the effects of the formation of the Modern State on the customs and morals of individuals.
3   According to the Universal Declaration of Human Rights (1948).
4   The term globalisation, in this study, is used not only as a mere concept of economic integration but, following the line of Chesneaux (1995), also as a process that involves transformations in the meanings of intensification of communications, time-space, deterritorialisation, world integration, technical modernity and social reflexivity.

in several aspects that integrate with society. We admit that globalisation refers to all aspects of life and that, depending on how it is questioned and operationalised, it may or may not be beneficial. Therefore, globalisation "has to do with the construction, codification, and distribution of knowledge that turns into goods for sale" (Skovsmose, 2014, p. 130). This way, mathematical knowledge is involved as part of the foundations of society, making it necessary to question its position in this laborious civilising equation (Civiero and Bazzo, 2020).

In this context we delegate some power to critical mathematics education (CME) by considering that it can contribute to the formation of critical individuals by promoting reflection on this process. Bazzo (2019) considers us to experience a civilising equation whose variables need to be discussed in schools. Also, Civiero (2016) shows that CME is today the most developed approach to dealing with contemporary variables in mathematics classrooms. The author discusses the importance of these questions being part of the training of mathematics teachers.

Based on the exposed understandings, we advocate the construction of landscapes of investigation[5] in mathematics classrooms to provide students and teachers with the opportunity to investigate themes that may foster reflection on contemporary issues. We identified as a possibility for the development of landscapes of investigation the Scientific Initiation that forms part of the curriculum in high school, at the Instituto Federal Catarinense—Rio do Sul Campus, Santa Catarina, Brazil.[6] It is a space for the development and alteration of perspectives on scientific activity, which provides undergraduates with an initiation into research in basic and higher education. It is also a path to intellectual independence, creativity, curiosity and autonomy, and can be a means to sharpen the students' critical awareness. According to Freire (1974):

> the critical consciousness is characterised by delving into the interpretation of problems; substituting causal principles for magical

---

5 For a characterisation of "landscape of investigation", see Skovsmose (2001a).
6 The Federal Institutes of Education, Science and Technology (IF) in Brazil are institutions that offer professional and technological education at all levels and modalities, forming and qualifying citizens to act in the different sectors of the economy, with emphasis on socioeconomic local, regional and national development. The IFs are present in all Brazilian states, covering approximately 80% of the country's micro-regions. The Instituto Federal Catarinense (IFC) is part of the federal network and comprises 15 campuses and the rectory distributed in the state of Santa Catarina. They are institutions that guarantee public, free and quality education.

> explanations; testing one's "findings"; and by opening to review; by attempting to avoid distortion when perceiving problems, and preventing preconceived notions when analysing them; by refusing to transfer responsibility; rejecting passive positions; by offering sound argumentation; by practicing dialogue rather than controversy; by receiving well the new for reasons beyond mere novelty and by having the common sense not to reject the old just because it is old—by accepting what is valid in both the old and the new (p. 15).

As an example, we present here part of Civiero's research (2009). At each step of the experience, we use the students' speeches, interspersed with theory, to highlight the constitution of the landscape of investigation.

Finally, we show that the landscape of investigation leads students and teachers to experience actions based on mathematics. In this way, we realise the imbrications between mathematical knowledge and the contemporary variables of this complex civilising equation.

1. Scientific Initiation in High School

Both Scientific Initiation and CME are possible places for the discussion of contemporary issues, and they contribute to the critical education of the students. Scientific Initiation is a fundamental space for research. Therefore, we contend that research is:

> the search, the study, the knowledge, the explanation, and the understanding of the world that surrounds it, motivated by actions of the subject that makes science. This demonstrates that it is not enough to fulfil the requirements of the system, it is also necessary to reduce the gap among areas of knowledge and between the technical and the human (Oliveira 2017, p. 32, our translation).

In line with this notion of research, according to Bazin (1983) and Oliveira et al. (2013), Scientific Initiation is a path of intellectual independence. As a scientific activity, it does not occur outside a social context.

Hence, we understand Scientific Initiation as a collaborative space of authorship experience, as the "search for the understanding in which the human being lives" (Oliveira, 2017, p. 32). Its insertion in basic education is pertinent, as Scientific Initiation offers scientific and technological education that contributes to the formation of the individual by fostering curiosity, creativity, authorship, decision-making and interpretation of reality through an initiation into research.

Despite its importance, Scientific Initiation in high school, which is aimed at students aged between fifteen and eighteen, is recent in Brazil. According to Oliveira (2017), it can currently be classified into three modalities: as an institutional program (since 1986), as public policy (since 2003) and as a component of the curriculum (since 2001).

Regardless of the modality, Scientific Initiation in high school must be distanced from technical rationality. That is, it must not focus on crystallised techniques and methodologies, such as imitation, repetition and reproduction, because:

> high school Scientific Initiation can articulate and integrate diverse knowledge, theory and practice and teaching, research, and extension. Dialogicity, problematisation, critical reflection, and collaboration are the basis for the development of people's autonomy. Based on these potentials, Scientific Initiation in high school is not just a space for methodological learning or research initiation focussed on training researchers concerned with an object of study that is alien to reality, society, and the civilising process (Oliveira, 2017, p. 147, our translation).

Oliveira, Civiero and Bazzo (2019) uphold the idea of Scientific Initiation as a component of the curriculum, a space in which the project selected for transposition was developed. For the authors, Scientific Initiation in high school "is a possibility to deal with contemporary issues and bring knowledge from different areas to the student's reality and, therefore, to bring reflective and critical discussions" (p. 469).

We advocate this potential of the Scientific Initiation. Therefore, we have the modality of Scientific Initiation as a curriculum component at the basis of this study, as it is an environment that guides the permanent reconstruction of knowledge. This modality takes place at the Instituto Federal Catarinense (IFC), Rio do Sul Campus,[7] locus of this study.

2. Scientific Initiation as a Curriculum Component in High School

In 2001, the Scientific Initiation Project started at IFC—Rio do Sul Campus as a project that was part of the Diverse element of the high

---

[7] The IFC—Rio do Sul Campus has existed as a Federal Education Institution since 1994. Currently, it has three technical courses integrated into high school (Agriculture, Agroecology and Informatics), six higher education courses, two lato sensu graduation courses and one course after high school.

school curriculum matrix, with a two-hour workload per week. One or two teachers manage this workload, where students are offered training on epistemological foundations of science, as well as aspects of research methodology, and the writing of projects and reports.

The insertion of the Scientific Initiation Project into the curriculum allows for the production of knowledge and the articulation of different areas of knowledge, minimising the boundaries between the curriculum components (Scheller et al., 2015). Therefore, there are structural elements that permeate the organisation of Scientific Initiation from the beginning, as shown in the chart below.

1st Grade of High School	2nd Grade of High School
-Theoretical and epistemological study on science, demystifying taboos on what it means to be a scientist;	- Project execution;
	- Collection, analysis, verification and systematisation of data, and materialisation of results;
- Methodological guidelines (problematisation and means of seeking solutions);	- FETEC (Technological and Scientific Knowledge Fair) summary preparation;
- Choice of theme and preliminary project;	
	- Work socialisation at FETEC;
- Study of the scientific and methodological foundations for the execution of the project;	- Lattes Curriculum—relevance and data filling;
- Selection of advisor;	- Work seminar concluded;
- Beginning of the project's development with supervision (justification, problem, first methodological steps, theoretical foundation);	- "Tutoring" of 1st-grade teams.
- Project elaboration and writing to be developed.	

Fig. 1. Structuring elements of the curriculum component Scientific Initiation. High school, IFC, Rio do Sul campus, 2001–2019.

With a view to the organisation in Figure 1, we realise that, at first, the teaching plans support the discussion of topics that instigate students' critical reflection on the world, leading them to perceive themselves as subjects in and of the world. Through talks and deconstructions of myths and taboos about science, technology and scientists, students start the project, which involves choosing a topic under the guidance of a campus professor.

Scientific Initiation as a curriculum component allows all students to participate, to establish dialogical teaching and learning relationships in a process of dodiscence. "Teaching, learning, and researching deal with these two moments of the gnoseological cycle: the one in which the existing knowledge is taught and learned, and the one in which the production of knowledge that does not exist is yet to be worked" (Freire, 1996, p. 28).

Thus, Scientific Initiation is a space aimed at instigating the student's enjoyment of—and inclination to—learn. In short, it is a place that may encourage students to be curious, reflective, to argue, to seek answers and to carry out the process of building knowledge in a critical way, which is not a tradition in education and society. In other words, Scientific Initiation is:

> an educational process capable of equipping the student for critical reading of the social practice in which he lives is the means that will make the school democratic. I understand that a democratic school takes the student to be a transforming subject of their reality; a critical look is not enough; the student must be inserted into the project, think, and plan necessary changes, believe in them, and put them into action. For this process to become effective, it is necessary to assume democratic attitudes when restructuring the didactic procedures (Civiero, 2009, p. 53, our translation).

This educational process that constitutes Scientific Initiation approaches the landscapes of investigation in problematisation and knowledge production. Besides, Scientific Initiation intertwines with scientific and technological education; therefore, it is a space for discussion about the variables of the civilising equation, so that the practice of Scientific Initiation:

> [...] needs, in its conduction and supervision process, a dialogical practice that problematises, that questions, that criticises knowledge, that values the other, that integrates, that instigates autonomy, and that takes care of life as the greatest social good, being essential the training of guiding teachers and/or researchers. The understanding and practice of scientific initiation need to go beyond the reproduction only

of issues already posed "culturally" for research and teaching, such as, for example, bureaucracy, elitisation, selective character, training, focus on the method, and reproduction of technical rationality. To impact upon humanising formation, it is necessary to have as a main pact the critical and reflective search to understand the world in which we live, established by a collaborative environment permeated by problematising dialogicity that relates science and technology and the civilising process, the advisor and the student (Oliveira, 2017, p. 275–276, our translation).

In this way, we recognise that the Scientific Initiation works developed in the IFC—Rio do Sul Campus are landscapes that encourage authorship and critical reflection on knowledge. Thus, it can be transposed into the classroom and constitute landscapes of investigation. Therefore, the first author selected Scientific Initiation projects and leveraged them into proposals of landscapes of investigation in mathematics classes. The author has reported this experience in her Master's dissertation (Civiero, 2009). She considers this a Reflective Didactical Transposition (RDT) process. The RDT process considers Didactic Transposition (Chevallard, 1991), but criticises and adds to this theory the concern about reflection on mathematical contents and the real context. In other words, the knowledge to be taught is adapted to provoke reflections on reality. Thus, RDT intends to transpose the knowledge to be taught in a landscape for investigation in order to instigate the subject's questioning of their own reality. In the process, student participation is promoted, and discussions and decision-making are provoked. Thus, the knowledge developed in the Scientific Initiation Project was transposed to mathematics classes, according to the CME perspective. For RDT, "it is not enough to transfer knowledge; it is necessary to instigate reflections on mathematical subjects linked to reality" (Civiero, 2009, p. 49). Below, we present the experience lived in one of the landscapes.[8]

3. Experiencing a Landscape of Investigation

Landscapes of investigation are milieus of learning built in the classroom. They allow for investigation, where students are invited to

---

8   Another possibility of RDT learning guides can be found in Civiero and Sant'Ana (2013).

make discoveries in a process full of questions, curiosities, explanations of perspectives and critical reflection. Therefore,

> the important point is that the landscapes of investigation are not explored based on a previous list of exercises. On the contrary, explorations take place through a "learning guide," in which students can point out directions, ask questions, ask for help, make decisions, etc (Skovsmose, 2001b, p. 64).

In this context, we carried out the RDT with two classes of first-grade high-school students at the IFC—Rio do Sul Campus. When developing the activity, reactions and comments from the students were observed, as well as the teacher's perceptions. The students' names have been substituted with letters of the Roman alphabet to conceal their identities.

In a context of uncertainty, in a landscape of investigation, it is paramount that the students accept the invitation to participate. According to Skovsmose (2001), a landscape of investigation is constituted when students accept (and assume themselves as active participants in) the process of exploration and explanation.

> [A] landscape of investigation is one that invites students to ask questions and seek explanations. The invitation is symbolised by its "Yes, what happens if ...?" In this way, students get involved in the exploration process. The teacher question: "Why is this?" represents a challenge, and the students' "Yes, why is this...?" indicates that they are facing the challenge, and that they are looking for explanations (Skovsmose, 2008, p. 21).

Therefore, we sought to encourage the investigation to arouse students' curiosity. We presented the material attractively, so that the activity was not seen as a command but rather as a different way to learn mathematics, to prioritise the reality in which the students were inserted. In this way, the teacher's role in this space is that of an inquirer and mediator, to avoid defined and unquestionable concepts.

At first, students were invited to investigate windrow behaviour through composting, based on the work "Composting From Various Organic Wastes", developed in the Scientific Initiation Project (2006/2007). The windrows are structured with a base of vegetable dry matter and layers are interspersed with organic matter. The system works with passive aeration, ensuring the thermophilic composting process. Figures 2 and 3 below illustrate the windrows.

The students who attended the Technical Course in Agroecology or Agriculture integrated with high school[9] showed interest. Thus, raising concerns about the environment as one of the contemporary variables in this globalised world is part of humanising education, as well as professional education.

4. From the Landscape to the Investigation

In the first stage of transposition (RDT) from the elements of work developed in the Scientific Initiation Project, students reflected on the importance of the project and the social relevance inherent to the theme. They also raised some hypotheses about its implementation. According to the authors of the Scientific Initiation work,

> the compost is very important for agriculture because it follows the concepts of agroecology, as it is a way of nourishing plants with all the macros and micronutrients that it needs without any type of external inputs that can harm nature (Battisti, Campos, and Souza, 2007, our translation).

The students began asking a series of questions that involved the subject: "What is organic matter? What is compost? What are the benefits of composting in the soil?" Such questions prompted the search for explanations. At the same time, students began to ask new questions: "How do you make compost?" (Student C). "What types of materials can you use to make the compost?" (Student J). "How do I prepare a compost pile?" (Student A). "How do I maintain this compost?" (Student P). "How do you check the maturity of the compost?" (Student K). "What are the stages of composting?" (Student B). In this way, the initial curiosity, which can be called naïve (Freire, 1996), was instated as they accepted the invitation.

However, some students opposed the activity, claiming that they did not perceive a relationship between the activity and mathematics. This resistance may be due to the exercise paradigm. Student C's question, "What does this have to do with the math class?" is an example of such an objection. They felt a little uncomfortable as they had to move from

---

9   An integrated course, it integrates high school with the technical course, which results in a single certificate of completion.

a passive to an active position. Resistance strengthened the premise of the ideology of certainty, imbued with a culture that supports the power of containing the ultimate argument attributed to mathematics. For Borba and Skovsmose (1997, p. 133), "students should, therefore, be persuaded against ideas such as a mathematical argument is the end of the story; a mathematical argument is superior by its very nature; the numbers say this and this." The only methodological option was to try to detach from the exercise paradigm. We started to discuss how urgent it was to use a different approach with the mathematical contents in order to analyse that reality.

After reading the proposition, the students delved into the theoretical foundations and were free to research more elements, which led them to become collaboratively involved. Classes started to gain momentum; students were not waiting for ready-made answers and investigated them when necessary, manifesting the criticism of the initial curiosity (Freire, 1996, 2006). Throughout the activity, the teacher felt she was in the risk zone, because, in a landscape of investigation, uncertainties are part of the process. Uncertainties also manifested in the students: "Teacher, when will mathematics appear?" (Student A) or: "This class is different, what does the teacher want?" (Student J).

After that, students were motivated to recognise the materials and methods used by the Scientific Initiation Project in the process. According to Figure 2:

Materials used: Cattle manure; bird dung with a small percentage of wood shavings (wood dust); raw kitchen waste; cooked kitchen waste and fibrous material (chopped elephant grass).

Procedures: Five windrows were made 90 cm long and 50 cm wide, each with a different type of organic waste, except for the fibrous material that was present in all windrows. We started the compost with a grass base, 10 cm high, 50 cm wide and 90 cm long. On the same day, we moved to the second layer using the other residues (raw kitchen waste, cooked kitchen waste, bovine manure, poultry manure with wood powder) each row with a type of waste 2 cm high. In the third layer, 10 cm of fibrous material was again placed in the fourth 2 cm of waste, and we ended with a fifth layer of grass. Right after finishing each 36 cm high windrow, we watered the compost with 6 litres of water. Finally, we cleaned the sides with a hoe.

Fig. 2. Adapted from Battisti et al. (2007).

Photos of the experiment were used to detail each step, to facilitate the understanding of the practice (Fig. 3 and Fig. 4).

Fig. 3. Taken from Battisti, Campos and Souza (2007).

Fig. 4. Taken from Battisti, Campos and Souza (2007).

After getting to know the compost production process, the next step was to start data analysis. At first, the teacher prompted the students to make estimates: "How do you think the composting occurred? How did the windrows behave? Which decomposed more quickly? How did this happen?" After many conjectures, Table 1 was created with the data.

Table 1. Behaviour of windrows with different residues for composting, Rio do Sul. Taken from Battisti et al. (2007).

Dates	Windrows of Other Types		Fibrous Widrow	
	Height Variation per Week (cm)	Final Height per Week (cm)	Height Variation per Week (cm)	Final Height per Week (cm)
09/11/06	0	30	0	30
16/11/06	-1.2	28.8	-2.4	27.6
23/11/06	-1.2	27.6	-3.6	24.0
30/11/06	-1.2	26.4	-1.2	22.8
07/12/06	-1.2	25.2	-2.4	20.4
14/12/06	-1.2	24	-2.4	18.0
21/12/06	-1.2	22.8	-1.8	16.8
28/12/06	-1.2	21.6	-1.8	15.0
05/01/07	-1.2	20.4	-0.9	14.1
12/01/07	-1.2	19.2	-0.7	13.4
19/01/07	-1.2	18	-0.5	12.9
26/01/07	-1.2	16.8	-0.7	12.2
03/02/07	-1.2	15.6	-1.2	11.0
10/02/07	-1.2	14.4	0	11.0
17/02/07	-1.2	13.2	0	11.0
24/02/07	-1.2	12	0	11.0

Based on the data, students were encouraged to analyse the results. Again, the teacher acted as an inquirer, triggering questions such as: "What data are listed in the title of the chart? Does the title tell us what is being presented? Tell us when and where the experiment took place. What can you see in the height variation of the windrows? What was the starting height?"

During the discussions, these observations emerged: "Look at the windmill with organic material, it always reduces equally. Did this really happen, teacher?" (Student C).

We emphasise that the way communication develops between students and the teacher can influence the learning (Alrø and

Skovsmose, 2002). The teacher plays a vital role in supervising this dialogue, as discussed by Milani (2017) and Milani et al. (2017).

In this process, the teacher instigated the students' understanding of the behaviour of the windrows, aiming to show them the difference between the windrows and the organic material (waste) and the fibrous windrow (straw).

> The windrow of fibrous material decreased faster due to disruption and varying the decrease, while the others gradually decreased by 1.2 cm per week. The windrows were made alternately with fibrous material and organic waste, and at the end, each windrow was 30 cm high. The different windrows gradually lowered around 1.2 cm each week, and in February, their height stabilised at 12 cm. The windrow that had only straw fell faster, varying its decrease due to decomposition and de-structuring of the compost, but ended up the same height as the others (Civiero, 2009, p. 83, our translation).

Student A observed: "Now I am aware of mathematics that appears in this data." Following this speech, other students spoke, willing to establish mathematical relationships and to represent them graphically. The students interacted and wanted to be involved in the discussions. The commitment and willingness were different from most maths classes the teacher had experienced before.

5. Mathematics-Based Actions

Figure 5 presents the behaviour of the data graphically, but also gives the mathematical models that have best adapted to the data.

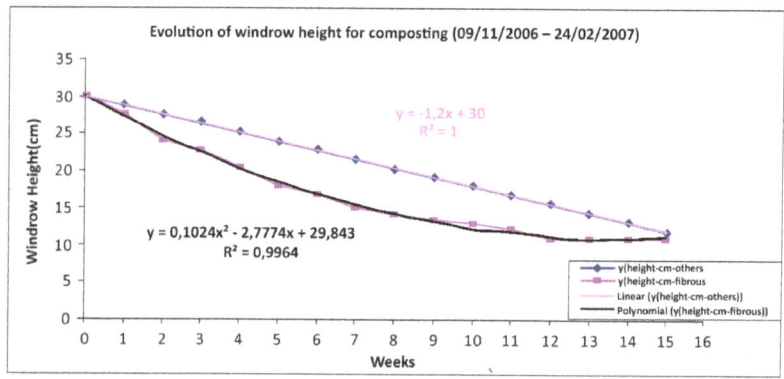

Fig. 5. Evolution of windrow height for composting. Taken from Battisti, Campos and Souza (2007).

This stage of RDT is in line with Skovsmose (2001b) regarding the three aspects of teaching and learning that the social argument of democratisation must present.

> 1) The material has to do with a real mathematical model; 2) The model has to do with important social activities in society; 3) The material develops an understanding of the mathematical content of the model, but this more technical knowledge is not a goal. The goal is to develop an insight into the hypotheses integrated into the model and thus develop an understanding of the processes (for example, decision processes) in society (Skovsmose, 2001b, pp. 43–44).

The developed activity converges with the three aspects. As for the latter, it was necessary to instigate discussions on the importance of models in a highly technological society. When analysing the model and its relationships, it is essential to check estimates and approximations. This allows the identification of an object of reflective knowledge, distinct from the object of technological knowledge.

The teacher asked questions: "What do they represent? What kind of curves appeared? What do the coefficients mean in the function?" Students were asked to define the mathematical model that best adapted to the curve.

Thereby, students became curious about mathematical models and started a process of mathematical discovery. In this case, specific mathematical knowledge and its concepts were essential to explain the reality. This speech can express some of the reactions: "Now I understand where the teacher wants to go. The project is full of mathematics" (Student D).

The students realised that in the windrow formed by various residues, the height variation was gradual, exactly 1.2 cm per week. They soon identified this number as the angular coefficient of the linear function, advancing to the concept of rate of change, which is made explicit in the statements: "The windrow is always lowering equally" (Student G). "Look at the table; each week, the windrow lowered 1.2 cm, it's the same number that appears in the function" (Student A).

Student B reflected: "This number is constant in the organic material windrow." Student J added: "Ah! That is why the graph is a line." Student D immediately asked about the meaning of the correlation index: "Teacher, what does this R^2 represent?" Students were encouraged to investigate this issue.

Meanwhile, Student C asked: "What about the windrow made of straw? It is different, so how do we identify the rate of change if it was not constant?" They realised that these windrows' behaviour was different, with the points not adapting to the linear shape, which would need a different mathematical analysis.

The students were curious about the functions that the Excel software presented. Student E asked: "Teacher, what do the charts have in common with these functions that Excel listed?" This question encouraged other students to speak out of curiosity. They were ready to start another stage of the investigation.

First, they understood that the Excel program had adjusted the curve according to the data that related the two quantities—that is, a set of coordinates (x, y). When asked which quantities were being related, Student A replied: "Of course, we are relating time (weeks) and height (cm)." Therefore, the variables x and y were revealed: variable x was the weekly variation, and the variable y was the windrow height variation. The students studied the mathematical concepts with interest and asked questions such as: What is that for? In traditional classes, this would usually become obsolete because this assumption had already been established. This issue disappeared as mathematical concepts were developed from the need generated in the context the students were inserted into.

In this phase of the explanation, we needed to refer to pure mathematics. Therefore, the students stopped to appropriate specific mathematical knowledge in order to understand the project. For the first windrow, which showed a linear decrease, they needed to study the function of first degree, recognising its main characteristics and rules. To exercise the content of linear systems, which emerged from the need to adjust curves, they used activities referring to semi-reality. Such activities make up an important part of the list of educational possibilities; however,

> [s]olving exercises referring to a semi-reality is a very complex competence and based on a well-specified contract between the teacher and the students. Some of the principles of this agreement are as follows: the exercise's wording fully describes semi-reality; no other information is essential for solving the exercise; more information is totally irrelevant; the only purpose of presenting the exercise is to solve it (Skovsmose, 2008, p. 25).

When carrying out the activities proposed, some students showed satisfaction, which they expressed in some comments: "Now, yes. The class became mathematics again" (Student C), which means that the student may still be framed by the exercise paradigm. However, this feeling was not shared by others, who said: "Wow, we did it again without knowing for what" (Student H). This student showed that they were uncomfortable with decontextualised exercises. And finally: "Of course not. We need to learn to calculate to understand the windrows" (Student M). We understand that this student realised that specific mathematical knowledge is needed to interpret reality. The student's initial curiosity was criticised. But then, as they deepened their knowledge, it became an epistemological curiosity (Freire, 1996).

During this traditional class, the students behaved formally, i.e., they reproduced the activities. However, this stage was full of meaning, and the students had a goal. Even so, the teacher needed to create space for dialogue, through active listening. According to Milani, Civiero, Soares, and Lima (2017, p. 240): "When the teacher tries to perform active listening, he starts the movement of dialogue that seeks to understand what the student says. This movement is not simple and immediate, as it is a change of posture, in the epistemological, methodological, and political sense."

In this process, it is possible to perceive students' involvement in different milieus of learning,[10] which highlights the potential of the landscape developed. Thus, they move between different milieus all the time, according to necessity. References to pure mathematics, in the context presented, are totally related to semi-reality and real-life. The landscapes of investigation concerning real life emerge naturally from a contemporary variable. Thus, this merging of milieus is not forced, like the grafts, but is a requirement for its development.

At the beginning of the following class, students were asked: "What did you see when observing the behaviour of elephant grass windrows and other waste? Why was it represented graphically by a line? How can you relate the data and the linear function presented by the Excel program?" After discussing, the students realised that the windrow height varies according to the passage of time (weeks). That is, in

---

10 The milieus were named according to the matrix elaborated by Skovsmose (2001a).

the function, it accompanies the x, which is representing the time in weeks. They also concluded that the parameter b of the function was representing the initial windrow height. To conclude, much dialogue was needed. Students went on debating, and, from time to time, the teacher intervened with a challenge to help them understand the mathematical relationships. The definitions became evident in students' statements: "Well, if the table shows that the windrow has always lowered by 1.2 and that each collection was made weekly, then, just by multiplying the week by 1.2, and we will know the windrow height" (Student H). "However, you can't forget that it has to be 1.2, because the windrow is lowering and a function $y = -1.2x + 30$ that expresses the windrow height according to the weeks is obtained" (Student C). "Interesting, then, that is why the graph is a straight line, week after week, the windrow decreases equally" (Student J). "Yes, the variation is always the same. I also noticed that the line is decreasing, which is logical because the windrow is lowering" (Student B). "Teacher, number thirty appeared in the function, is it the initial windrow height, or just a coincidence?" (Student G). "It is obvious, you see, now I understand. The graph is showing exactly the behaviour of the windrow. It starts with 30 cm and lowers by 1.2 cm, which must be represented by 1.2 cm because it is going down, i.e., it is decreasing" (Student D).

During the dialogue, there was active listening, and the students tried to emphasise the appropriate nomenclature. For example, when they said that the windrow was lowering, they were taught to use the term "decreasing", which was related to the position of the line. Gradually, they adapted language and mathematical symbology.

6. Different Approaches

Next, the students chose their resolution method: either the Simple Linear Regression or the Least Square Method. They found the values of the coefficients a and b, whose meanings and values they had previously recognised, and checked the mathematical algorithm. The variables x and y were already known. For example,

$$y = ax + b, \text{wherein:}$$

$$a = \frac{\left[n\sum(xy) - \left(\sum x\right)\left(\sum y\right)\right]}{\left[n\sum x^2 - \left(\sum x\right)^2\right]} \qquad a = \frac{-6528}{5440} = -1,2$$

$$b = \frac{\left(\sum y - a\sum x\right)}{n} \qquad b = \frac{480}{16} = 30$$

They used different methods to solve the problem. The purpose of fostering the study of different methods to determine parameters a and b is consistent with their desire to create decision situations. The students can decide which path they want to follow when developing the work. This freedom to choose the method also occurs in Scientific Initiation projects.

Knowing the parameters, we proceeded to interpret the model. For this, we had to observe the usual correlation coefficient and the determination coefficient. After an investigation of the mathematical devices needed to solve this situation, the students used the correlation coefficient formula:

$$r = \frac{n\sum(xy) - (\sum x)(\sum y)}{\sqrt{n\sum x^2 - (\sum x)^2} \cdot \sqrt{n\sum y^2 - (\sum y)^2}}$$

This calculation requires students to understand that a perfect correlation is one that approaches 1, which shows a perfect match between the data and the function. Student K interrupted, saying: "So, the result is always 1?" To answer the question, we argued that "the ideal would be 1, which would indicate a 100% correlation. However, this is difficult to achieve in practice because many factors determine the relationships between variables in real life. In our case, it did occur because the variation was constant, and all points were aligned, but it is not always so. Usually, some points are off the line, so the correlation index will not be exactly 1."

After the discussions, we suggested that students apply the formula to understand the concepts discussed. Then, replacing the values obtained before, they found the following calculations:

$$r = \frac{16.2112 - 120.336}{\sqrt{16.1240 - 14400} \cdot \sqrt{16.7545,6 - 112896}} = -1$$

When they saw a negative result, there was some anxiety. Student G immediately exclaimed: "It's wrong, look, it is negative!" Then, we discussed what this negative result would mean and that it would be interesting to explore its concept before excluding it. For this, they found that the coefficient can vary from -1 to 1, indicating that there is a strong relationship between the variables. What is explicit in the model as weeks go by, (x), the windrow is decreasing. That is, it is decomposing and, thus, reducing in size, which also happens with the graph. In the same way, we proceeded to the calculation and analysis of r^2 and the rate of change.

At this point, it is possible to infer that this dynamic occurred according to Skovsmose (2008, p. 13), who considers "that a new critical mathematical education must seek educational possibilities (and not propagate ready-made answers)". Postman and Weigartner (1969), who discuss the importance of changing attitudes, moving from a school of answers to that of questions.

7. The Model: The Importance of Critical Reflection

Asked to write a report on the main mathematical characteristics evidenced in the investigation, the teacher drew attention to one more detail: "Can we consider the function indefinitely?" Soon, Student C said: "It is absurd!", along with a colleague, Student J who, until then, had not spoken but was aware of everything: "It stopped decomposing." Student A also intervened: "If we only consider the function, the windrow height will begin to decrease indefinitely after the fifteenth week, which is absurd, because the height of the windrow has stagnated." Student G then expressed his curiosity: "And now, what do we do? How does mathematics explain this?"

After they spoke, we explained that it is necessary to indicate the domain of this function. Some students realised that the image (y) was a consequence of the domain (x) and reached a conclusion, expressing the answers as follows:

$$D(x) = \{x \in R \mid 0 \leq x \leq 15\} \text{ and } Im = \{y \in R \mid 12 \leq y \leq 30\}$$

During the RDT development in classes, we also highlighted the discussion about the importance of adjusting the curves. Different actual situations can present problems that require solutions and decisions that can be solved by a mathematical formulation. Thus, a mathematical model is represented by symbols and mathematical relationships that seek to translate, in some way, a phenomenon of reality. In this perspective, when proposing a model, we must keep in mind that it comes from approximations made to try and understand a phenomenon better. Thus, these approaches are not always consistent with reality, but they portray aspects of the situation analysed. Therefore, it is necessary to criticise the model in all its dimensions.

We highlighted that several observations could be made regarding mathematics in action and thus justify how mathematics can operate in technologies, production, management schemes and decision-making. As part of the laborious civilising equation, it can change social and cultural behaviours.

We discussed how society is technologised and how much mathematics helps to shape this society, which overlaps with Skovsmose's (2008, p. 112) third concern about mathematics in action, when he states that it "is a paradigmatic space to discuss structures of knowledge and power in today's society."

In this debate, we emphasise the accelerated change in the civilising equation led by the Fourth Industrial Revolution. According to Schwab (2017), this behaviour announces a 4.0 Revolution, characterised by the transition towards new systems that overcome the digital revolution. Consequently, the power relations underlying the processes of commercialisation and industrialisation are closely linked to technoscientific development, which, in turn, is conditioned by mathematical algorithms.

We also observed that mathematics provides the possibility of hypothetical reasoning, i.e., it can analyse the consequences of an imaginary landscape. On the other hand, mathematics can also help to construct (true or false) justifications to legitimise some decisions and actions. The students raised the issue of the electoral season when we look at the polls. Depending on their organisation, they do not always represent reality but are used to influence and pressure us to choose a specific candidate. In 2020, we could discuss statistical data from the COVID-19 pandemic. For example, the case fatality rate (CFR) depends on the number of confirmed cases, and for that, it depends on the number of tests that are carried out. Therefore, it is extremely challenging to make accurate estimates of the actual death risk.

The students were frightened by the power of mathematisation, which could be seen by the speech: "Wow! Mathematics has power over everything; it gives me a shiver. Knowing that everything is calculated in advance, and callously" (Student J).

We tried to sharpen the discussion by emphasising that mathematical models are not always constructed from a socially just perspective. Encouraging critical mathematics is "integrating students' lives, knowledge, and cultures; having students learn important mathematics and about their world; and supporting them to act on the injustices they perceive and experience" (Gutstein, 2012, p. 65). Therefore, to paraphrase Skovsmose (2008, p. 118), we emphasised that "mathematics should be a theme for reflection and criticism in all its forms of action."

In this context, we sought to show how critical education can be guided toward emancipation. These discussions were based on Skovsmose (2008, p. 94), when he mentions critical citizenship: "[...] it can 'challenge' the constituted authority. It carries with it the opposition to any decision considered unquestionable." We highlighted the relevance of knowing how society is managed and how situations are planned, often mathematically, with its algorithms.

Consequently, this educational possibility was planned, aiming to bring about changes to problematise the need to criticise the social system. During the discussions, the significance of dialogue between the teacher and the students became evident. However, we agree with Skovsmose's concerns when he states that:

the important question now is how well mathematics education can prepare [students] for critical citizenship. I do not see that such preparation is related to the school's mathematical tradition. I do not even see it linked to the intimate nature of mathematics. It has to do with a possible function of mathematics education (Skovsmose, 2008, p. 95).

We argue that discussions of this type can help mathematics education set critical citizenship in motion. In this way, students can experience actions based on mathematics to realise how relevant reflections are. One of the students commented in between discussions in class: "We cannot accept everything as finished, full stop; we need to understand the process to be able to accept it or not" (Student F).

We concluded the study of the function of the first degree with this analysis. However, we ended the class by encouraging students to observe the behaviour of the windrow made up of fibrous material, whose model refers to a quadratic function. Nevertheless, we will talk about this development at another time, or it can be seen in Civiero (2009).

For Civiero and Sant'Ana (2013, p. 695), the landscapes of investigation from Scientific Initiation constitute a "critical reflexive approach that can relate teaching to the act of questioning and making decisions, establishing a link with life in society and mathematics."

In this context, we advocate the relevance of mathematics teachers' scientific and technological literacy, so that they are prepared to promote RDT through landscapes of investigation about contemporary variables. In this regard, Civiero, Fronza, Oliveira, Schwertl and Bazzo (2017, p. 2673), state that:

> for the teacher to develop mathematical concepts related to reality, he needs to recognise it, read the news critically, expand his list of readings, understand, make decisions, evaluate and criticise social, political, economic, scientific, and technological issues. Recognise the dynamics and complexity of the educational world imbricated with reality outside it, acting cooperatively and collaboratively (our translation).

In this way, we emphasise that the teacher must assume a critical epistemological conception to foster a critical posture in his students, making them capable of analysing and making decisions that can interfere in reality and, consequently, in people's quality of life.

8. Additional Considerations

Hamlet's dilemma: "To be or not to be, that is the question", one of the most famous phrases in world literature, may seem complex, but it is actually very simple. "To be or not to be" is about acting, decision-making and positioning oneself or not in the face of events. In this perspective, the phrase led us to question the possibilities of CME in today's society.

We live in a civilising equation on a different scale, scope and complexity from any that has ever occurred before. The challenges for those who aspire to social justice because of the equation are increasingly complex. Therefore, we must urgently enter all possible spaces to show that a better world is possible, a world where all people have equal and inalienable rights as the foundation of freedom, justice, peace and social development.

For this purpose, the description of the Reflective Didactic Transposition (RDT) of a Scientific Initiation project for mathematics classes provided the development of a landscape of investigation, which, in turn, encouraged reflection and criticism. The development of this approach evidenced the approximation of Scientific Initiation conceptions with the landscapes of investigation. Therefore, when proposing activities from the perspective of the CME, we provide discussions about the variables of the civilising equation. It is a way to equate the many elements of the interwoven relationship between technical aspects and human issues.

When exploring the behaviour of the composting windrows (the theme of the Scientific Initiation Project), our objective can be an environmental study, one of the essential variables that reflect fundamentally on the existence of the Earth. We invited students to get involved in the production of the mathematical model and reflect on how the results are related to the criteria used and how they can be used in society. We emphasise that mathematics plays an essential role in social issues. Therefore, we reiterate the importance of looking at the context and appropriating it for reflection and action to foster collective decision-making, given the needs established in the process.

It was also possible to highlight the need for urgent imbrication of the variables, the scientific initiation, and the different milieus of learning. When looking for different milieus, one can articulate the dimensions of the methodological specificities of pure mathematics, semi-reality and reality. In this context, the landscapes of investigation with real-life references—environmental variables related to the professional course—took shape. Its content was essential for the interpretation of life through mathematics, which is found in Skovsmose's (2005, p. 96) words: "In this way, they experienced what actions based on Mathematics can mean and realized the importance of reflection."

Finally, we argue that the development of landscapes of investigation is fundamental to highlight mathematics linked to technological and human issues that constitute the civilising equation. To this end, it is essential to provide discussions inherent to reality to appropriate specific mathematical knowledge imbricated with other areas of knowledge, therefore applying them to promote a society where the principles of human dignity are guaranteed, and social justice prevails.

References

Alrø, H. & Skovsmose, O. (2002). *Dialogue and learning in mathematics education: Intention, reflection, critique.* Kluwer Academic Publishers

Battisti, G., Campos P. C. de & Souza, W. O. (2007). Compostagem a partir de diversos resíduos orgânicos. *Relatório de iniciação científica.* EAFRS: Rio do Sul.

Bazin, M. J. (1983). O que é Iniciação Científica. *Revista do Ensino de Física, 5*(1), 81–88.

Bazzo, W. A. (2016). Ponto de ruptura civilizatória: A pertinência de uma educação "desobediente". *Revista CTS, 33*(11), 73–91.

Bazzo, W. A. (2019). *De técnico e de humano: questões contemporâneas.* Ed. da UFSC.

Borba, M. & Skovsmose, O. (1997). The ideology of certainty. *For the Learning of Mathematics, 17*(3), 17–23.

Chevallard, Y. (1991). *La Transposition didatique: du savoir savant au savoir enseigné.* La Pensée Sauvage.

Civiero, P. A. G. (2009). *Transposição Didática Reflexiva.* Master's thesis. Universidade Federal do Rio Grande do Sul, Porto Alegre.

Civiero, P. A. G. & Sant'Ana, M. F. (2013). Learning roadmaps from the reflexive Didactic Transposition. *Bolema, 27*(46), 681–696.

Civiero, P. A. G. (2016). *Educação matemática crítica e as implicações sociais da ciência e da tecnologia no processo civilizatório contemporâneo: Embates para formação de professores de matemática.* Doctoral dissertation. Universidade Federal de Santa Catarina, Florianópolis.

Civiero, P. A. G., Fronza, K. R. K., Oliveira, F. P. Z., Schwertl, S. L. & Bazzo, W. A. (2017). Alfabetização Científica e Tecnológica no Currículo da Formação de Professores de Matemática. *Ensenanza de Las Ciencias, Extra,* 2669–2674.

Civiero, P. A. G. & Bazzo, W. A. (2020). A equação civilizatória e a pertinência de uma educação insubordinada. *International Journal for Research in Mathematics Education, 10*(1), 76–94.

Civiero, P. A. G & Oliveira, F. P. Z. de. (2020). Landscapes of investigation and scientific initiation: Possibilities in civilizatory Equation. *Acta Scientiae (Canoas), 22*(5), 165–185.

Freire, P. (1974). *Education for critical consciousness.* British Library.

Freire, P. (1996). *Pedagogia da autonomia: saberes necessários à prática educativa.* Paz e Terra (Coleção Leitura).

Freire, P. (2006). *À sombra desta mangueira.* Olho dágua.

Gutstein, E. (2012). Reflections on teaching and learning mathematics for social justice in urban schools. In: *Teaching mathematics for social justice: Conversations with educators.* Reston: NCTM, 63–78.

Milani, R. (2017). "Sim, eu ouvi o que eles disseram": o diálogo como movimento de ir até onde o outro está. *Bolema, 31*(57), 35–52.

Milani, R., Civiero, P. A. G., Soares, D. A. & Lima, A. S. de. (2017). O Diálogo nos Ambientes de Aprendizagem nas Aulas de Matemática. *RPEM, Campo Mourão, Pr. 6*(12), 221–245.

Oliveira, F. P. Z., Civiero, P. A. G., Fronza, K. R. K., & Mulinari, G. (2013). Iniciação Científica para Quê? *Enseñanza de las Ciencias, Extra,* 2764–2768.

Oliveira, F. P. Z. de. (2017). *Pactos e impactos da Iniciação Científica na formação dos estudantes do Ensino Médio.* Doctoral dissertation. Universidade Federal de Santa Catarina, Florianópolis.

Oliveira, F. P. Z., Civiero, P. A. G & Bazzo, W. A. (2019). A iniciação científica na formação de estudantes do ensino médio. *Revista Debates em Educação, 11*(24), 453–473.

Postman, N. & Weingartner, C. (1969). *Teaching as a subversive activity.* Delacorte Press.

Scheller, M., Civiero, P. A. G. & Oliveira, F. P. Z. (2015). Pedagogical actions of reflective mathematical modelling. In G. A. Stillman, W. Blum & M. S. Biembengut (Eds), *Mathematical modelling in education research and practice: Cultural, social and cognitive influences* (pp. 397–406). Springer.

Schwab, K. (2017). *The fourth industrial revolution.* Crown Business.

Shakespeare, W. (2009). *Hamlet.* Cambridge University Press.

Skovsmose, O. (2001a). Landscapes of investigation. *ZDM—Mathematics Education, 33*(4), 123–132.

Skovsmose, O. (2001b). *Educação matemática crítica: A questão da democracia*. Papirus.

Skovsmose, O. (2005). *Travelling through education: Uncertainty, mathematics, responsability*. Sense Publishers.

Skovsmose, O. (2008). *Desafios da reflexão em educação matemática crítica*. Papirus.

Skovsmose, O. (2014). *Critique as uncertainty*. Information Age Publishing.

Contributor Biographies

Mario Sánchez Aguilar is from Mexico. He is interested in political aspects of mathematics education research and practice. He is head of the Mathematics Education Program of the National Polytechnic Institute of Mexico. Currently he serves as Associate Editor of the Ibero-American research journal *Educación Matemática*. He is a visiting professor at the University of San Carlos of Guatemala. He has a bachelor's in mathematics from the University of Guadalajara, Mexico (1999), and a Master's degree in mathematics education from CINVESTAV-IPN, Mexico, 2003. He achieved his PhD in mathematics education at Roskilde University, Denmark (2010).
Orcid: http://orcid.org/0000-0002-1391-9388
Email: mosanchez@ipn.mx

Bülent Avcı lives in the USA. His research focusses on dialogical teaching of high school mathematics, critical pedagogy, and critical mathematics education oriented towards critical citizenship. He teaches at high school and university level in the USA. He is the author of the book, *Critical Mathematics Education: Can Democratic Mathematics Education Survive under Neoliberal Regime*. He publishes articles both in peer-reviewed and non-peer-reviewed journals. Bülent Avcı is the editor of peer-reviewed online journal, *Rethinking Critical Pedagogy*.
Orcid: https://orcid.org/0000-0002-0544-5899
Curriculum: https://scholar.google.com/citations?user=6zKsVCQAAAAJ&hl=en
Email: mjura41@hotmail.com

Denner Dias Barros is a PhD student at the Graduate Program in Mathematics Education, Universidade Estadual Paulista (Unesp), Campus Rio Claro, São Paulo. He has a Master's degree in mathematics education from Unesp (2017). He graduated in mathematics from Unesp

(2015), and in pedagogy from Faculdade Educacional da Lapa (FAEL), (2018). He is a specialist in Brazilian Sign Language. His interests are inclusive education, LGBT + communities, gender, and sexuality as part of a preoccupation of critical mathematics education.
Orcid https://orcid.org/0000-0002-8108-022X
Curriculum: http://lattes.cnpq.br/7180580652216737
Email: dennerdias12@gmail.com

Lucicleide Bezerra is a PhD student in mathematics and technological education at Universidade Federal de Pernambuco (UFPE). She is investigating the teaching and learning of statistics based on critical mathematics education. In 2014, she obtained a Master's degree in mathematics and technology education for the same university. In 2002, she obtained a specialisation degree in administration from Faculdade Frassinetti do Recife (Fafire). In 1999, she achieved a bachelor's degree in statistics from Universidade Católica de Pernambuco (UNICAP). In 2016, she graduated in mathematics from Universidade Paulista (UNIP) in São Paulo. The focus of her research is the teaching and learning of statistics, statistical literacy, and critical mathematics education.
Orcid: https://orcid.org/0000-0003-2791-3398
Curriculum: http://lattes.cnpq.br/4376662679246462
Email: lucicleide.bezerra@gmail.com

Arindam Bose lives in India. He is a mathematics education researcher and curriculum developer. He is currently an Associate Professor in Tata Institute of Social Sciences (TISS), Mumbai, India and was a visiting professor (2019–2020) and currently is a collaborative professor at the Universidade Federal de São Paulo (Unifesp), Diadema, Brazil. His areas of interest include sociology of mathematics education, implications of out-of-school mathematics and funds of knowledge for school mathematics learning, critical mathematics education, history of mathematics education, multilingualism and mathematics learning.
Orcid: https://orcid.org/0000-0003-2209-2092
Curriculum: https://tiss.edu/view/9/employee/arindam-bose/
Email: arindambose.ab@gmail.com

Reginaldo Ramos de Britto is a mathematics teacher in basic education and in high school. In 2000, he specialised in science education. In 2012, he obtained a Master's degree in mathematics education, and in

2017 a bachelor´s in humanities, both from the Federal University of Juiz de Fora, Brazil. He is a PhD student at the Graduate Program in Education at the Universidade Federal de Juiz de Fora. He is also an undergraduate student in social sciences at the same university. He is a member of the Brazilian Association of Black Researchers (ABPN) in science and technology. His interests include decolonial studies, critical mathematics education, and afro-ethnomathematics.
Orcid: https://orcid.org/0000-0002-2957-7511
Curriculum: http://lattes.cnpq.br/0080602618573783
Email: reginaldorrbritto@gmail.com

Manuella Carrijo is a PhD student in mathematics education at the Universidade Estadual Paulista (Unesp), Campus Rio Claro, São Paulo. She is currently an intern at the University of Klagenfurt, Austria. She has a Master's degree in science and mathematics education and a bachelor's degree in mathematics, both from Universidade Federal de Goiás. Her research interests include critical mathematics education focussed on the following subjects: immigrant students, citizenship, critical race theory, mathematics education for social and racial justice, and evaluation in the teaching and learning of mathematics.
Orcid: https://orcid.org/0000-0002-5879-7652
Curriculum: http://lattes.cnpq.br/6353436161500379
Email: manuellaheloisa@gmail.com

Paula Andrea Grawieski Civiero's research interests are teacher education, critical mathematics education, and social implications of science and technology. She obtained a degree in mathematics (1994), and a Master's degree in mathematics education (2009). She obtained a PhD in science and technology education from the Universidade Federal de Santa Catarina (UFSC) (2016), and post-doc at UFSC (2021). She has been a researcher and lecturer at Instituto Federal Catarinense (IFC), Campus Rio do Sul, Santa Catarina, Brazil, since 2005. She works with the Mathematics Fairs Movement, with the Nucleus of Studies and Research in Technological Education (NEPET/UFSC). She is also Deputy Leader of the Center for Studies and Research on Mathematics Education (NEPEMP/IFC).
Orcid: http://orcid.org/0000-0002-5841-7330
Curriculum: http://lattes.cnpq.br/6617701172635064
Email: paula_civiero@yahoo.com.br

Ana Carolina Faustino's research interests include critical pedagogy, critical mathematics education, mathematics education for social justice, language and communication in the mathematics classroom, and mathematics education in the early years of elementary school. Since 2019, she has been Associate Professor in Education at Universidade Federal de Mato Grosso do Sul, Brazil. She has a bachelor's degree in pedagogy and a Master's degree in education, both from the Universidade Federal de São Carlos (Ufscar), Brazil. She has a PhD in mathematics education from the Universidade Estadual Paulista (Unesp), Brazil. She is a member of the Academic Advisory Board for the journal *Rethinking Critical Pedagogy*.
Orcid: https://orcid.org/0000-0002-2059-9466
Curriculum: http://lattes.cnpq.br/7789919599029387
Email: carola_loli@yahoo.com.br

Edyenis Frango graduated in mathematics from Universidade Federal Fluminense, Rio de Janeiro (2016). She has a Master's degree in mathematics education from Universidade Federal de Juiz de Fora (2019). She is member of the Pesquisa de Ponta research group and has as her main research interests mathematical modelling, teacher education from the perspective of critical mathematics education, and mathematics education for social justice. She is interested in inclusive mathematics education, the subject of her graduation thesis. She has teaching experience in basic education. Currently, she is Assistant Professor at Universidade Federal Fluminense, working in teacher education.
Orcid: https://orcid.org/0000-0001-7865-1135
Curriculum: http://lattes.cnpq.br/3555779492614250
Email: edyenisfrango@gmail.com

Íria Bonfim Gaviolli graduated in mathematics from the Universidade Estadual do Paraná (Unespar) in 2016. From 2012 to 2016, she worked as an elementary school teacher. In 2018, she obtained a Master's degree in mathematics education from the Universidade Estadual Paulista (Unesp), Campus Rio Claro, São Paulo. Her thesis is entitled "Landscapes of Investigation and Mathematics Education from a Perspective of Deficiencialism". Currently, she is a PhD student in mathematics education at the same university. Her research interests

are inclusion in mathematics education and philosophy of mathematics education, focussing on deficiencialism (the invention of disability by normality), inclusion as an imperative, inclusion/exclusion and Neoliberalism, production of subjectivities, and childhood and autism.
Orcid https://orcid.org/0000-0003-4751-3412
Curriculum: http://lattes.cnpq.br/6356093078287037
Email: iriagaviolli@gmail.com

Rejane Siqueira Julio is Associate Professor in Mathematics Education at Universidade Federal de Alfenas, Brazil, since 2009. In 2007, she obtained a Master's degree in mathematics education at Universidade Estadual Paulista (Unesp), Campus Rio Claro, São Paulo. In 2015, she completed her PhD study in education at Universidade Estadual de Campinas (Unicamp). Her research interests relate to teacher education and professional development from the perspective of the model of semantic fields and the philosophy of Ludwig Wittgenstein. She is interested in approximations between these perspectives and critical mathematics education.
Orcid: https://orcid.org/0000-0002-3248-800X
Curriculum: http://lattes.cnpq.br/1798884495942862
Email: resiju@gmail.com

Adriana de Souza Lima has a special interest in financial education research, focussed on teacher education from the perspective of critical mathematics education and decolonial thinking. She acquired broad professional experience in elementary school as a mathematics teacher in public schools in the state and city of Rio de Janeiro. She developed several works based on students in situations of social vulnerability, including mathematics and capoeira. In 2000, she obtained a degree in mathematics. In 2016, she obtained a Master's degree in elementary education. She is a PhD student in mathematics education at the Universidade Federal do Rio de Janeiro (UFRJ).
Orcid https://orcid.org/0000-0002-5970-362X
Curriculum: http://lattes.cnpq.br/6744865441596524
Email: adridlima@yahoo.com.br

Agustín Méndez is from Mexico. He is interested in the practice of mathematics education. He is a secondary and high school mathematics

teacher at Colegio Las Hayas de Coatepec, Veracruz, Mexico. He currently works as a collaborator in the "Mathematics for All" project from the Ministry of Education of Veracruz. He obtained a degree in mathematics from the University of Veracruz, Mexico, in 2008. In 2018, he obtained a Master's degree in mathematics education from CICATA-IPN, Mexico.
Orcid: https://orcid.org/0000-0002-2268-1949
Email: agustin@hayas.edu.mx

Raquel Milani is Associate Professor at the Universidade de São Paulo (USP), in Brazil. She develops research at the Graduate Program in Education—USP. Considering the perspective of critical mathematics education, her research interests include mathematics teacher education, the learning of dialogue, and mathematics learning and teaching. She has a teaching degree in mathematics (2000) from Universidade Federal do Rio Grande do Sul (UFRGS), and a Master's degree (2003) and PhD (2015) in Mathematics Education from Universidade Estadual Paulista (Unesp).
Orcid: https://orcid.org/0000-0002-2015-7641
Curriculum: http://lattes.cnpq.br/8977517401117545
Email: rmilani@usp.br

Yael Carolina Rodríguez Moreno is from Colombia. Her interest is in critical mathematics education and on relevant social situations that promote reconstruction of social structures. For fifteen years, she has worked in the field of mathematics in both private and public sector. During the last four years, she has been mentoring at the Education Ministry for the Teaching and Academic Excellence program called "Everyone to Learn", where she supports teachers and heads of educational centres from the Secretary of Education of Soacha, Cundinamarca, Colombia. As a part of her academic background, she has a degree in social work awarded by the Fundación Universitaria Claretiana. She also holds a degree in mathematics and a Master's degree in education, with emphasis on mathematics education, awarded by the Universidad Distrital Francisco José de Caldas. She has a degree in university teaching, awarded by the Universidad Militar Nueva Granada.
Orcid: https://orcid.org/0000-0001-6356-7483
Email: yaelcarolinarodrim@gmail.com

Amanda Queiroz Moura's research interests are critical mathematics education, landscapes of investigation, dialogue, and inclusion. She has developed research aimed at the organisation of teaching and learning environments in mathematics focussing on diversity and the search for equity and social justice. Currently, she is a senior researcher at Klagenfurt University and a research collaborator in the Épura Group (Unesp), which addresses a range of issues related to problems of social inclusion-exclusion, considering the perspective of critical mathematics education. She has a degree in mathematics (2011), a Master's degree (2015), and a PhD (2020) in mathematics education from Universidade Estadual Paulista (Unesp), Campus Rio Claro, São Paulo, Brazil.
Orcid: https://orcid.org/0000-0002-9472-3773
Curriculum: http://lattes.cnpq.br/3013921866070769
Email: amanda_qm@yahoo.com.br

Fátima Peres Zago de Oliveira's research interests are teacher education, critical mathematics education, and social implications of science and technology. She obtained a degree in mathematics in 1990, and a Master's degree in computer science in 2004. She has a PhD in science and technology education at Universidade Federal de Santa Catarina (UFSC) (2017). Since 1995, she has been a researcher and lecturer at Instituto Federal Catarinense (IFC), Campus Rio do Sul, Santa Catarina. She has been Vice President of the Sociedade Brasileira de Educação Matemática (SBEM) (2019–2022). She works with the Mathematics Fairs Movement, with the Nucleus of Studies and Research in Technology Education and with the Research Group on Science and Technology Education in Professional Education, IFC.
Orcid: http://orcid.org/0000-0002-9114-8611
Curriculum: http://lattes.cnpq.br/2928350526317146
Email: fatima.oliveira@ifc.edu.br

Miriam Godoy Penteado's research interests are teacher education and collaboration between universities and schools with a focus on inclusive mathematics education. Together with Ole Skovsmose, she coordinates the Épura research group—including researchers, doctoral students, Master's students, and teachers—addressing a range of issues related to social inclusion-exclusion, considering the perspective of critical mathematics education. She has a degree in mathematics, a Master's degree in mathematics education, and a PhD in education

at Universidade de Campinas (Unicamp). She completed her postdoc at the University of Bristol, and has been Associate Professor at Universidade Estadual Paulista (Unesp), Campus Rio Claro, São Paulo, Brazil, since 1989. Currently her main activity is associated with the Graduate Program in Mathematics Education at Unesp, as volunteer researcher.
Orcid: https://orcid.org/0000-0003-0458-275X
Curriculum: http://lattes.cnpq.br/4099992332439295
Email: miriam-godoy.penteado@unesp.br

Renato Douglas Ribeiro is currently a PhD student in mathematics education at Universidade Estadual Paulista (Unesp), Campus Rio Claro, São Paulo. He has a Master's degree in education from Universidade de São Paulo (USP), 2012. He graduated in mathematics from the Universidade de São Paulo (USP), 2006. Since 2014, he has been a lecturer at Instituto Federal de São Paulo (IFSP), Campus Caraguatatuba, São Paulo, where his main activity is teaching preservice and in-service mathematics teachers.
Orcid: https://orcid.org/0000-0003-4120-8994
Curriculum: http://lattes.cnpq.br/9672139184699299
Email: redouglas@gmail.com

Fanny Aseneth Gutiérrez Rodríguez is from Colombia. Her special interests are critical mathematics education, the democratisation of knowledge, and the possible contribution of mathematics to citizen education. She has developed research projects linked to the groups Mescud and EdUtopía at the Universidad Distrital Francisco José de Caldas, Bogotá, Colombia. These groups work with undergraduate mathematics teachers and school students. She has twelve years of experience as a teacher, and for the last six years she has worked as a mathematics teacher in public schools for the Secretary of Education of Bogotá. She is an author of mathematics schoolbooks for Latin America and Spain. She graduated as a mathematics teacher and obtained a Master's degree in education with special emphasis on mathematics—both degrees were awarded by the Universidad Distrital Francisco José de Caldas.
Orcid: https://orcid.org/0000-0003-3970-1421
Email: asenethgr@gmail.com

Guilherme Henrique Gomes da Silva has a research interest in mathematics education for social justice. In 2016, he concluded his PhD study addressing the role of mathematics education in the face of affirmative action policies in higher education. In 2017, his work received an award from the Coordenação de Aperfeiçoamento de Pessoal de Nível Superior—Brasil (CAPES), Ministry of Education in Brazil. Currently, he is concerned with understanding how mathematics education could contribute to the academic progress of affirmative action students from science, technology, engineering, and mathematics (STEM) programs in Brazilian higher education. Since 2012, he has been Associate Professor in Mathematics Education at the Universidade Federal de Alfenas.
Orcid: https://orcid.org/0000-0002-4166-2663
Curriculum: http://lattes.cnpq.br/5817829882396943
Email: guilhermehgs2@gmail.com

Ole Skovsmose has a special interest in critical mathematics education. He has investigated the notions of landscape of investigation, mathematics in action, students' foregrounds, and pedagogical imagination. He has been Professor at the Department of Education, Learning and Philosophy (now Department of Culture and Learning), Aalborg University, Denmark. Currently, he is associated with the Graduate Program in Mathematics Education at the Universidade Estadual Paulista (Unesp), Campus Rio Claro, São Paulo as volunteer researcher. Together with Miriam Godoy Penteado, he coordinates the Épura research group at Unesp. He has published several books including *Towards a Philosophy of Critical Mathematics Education, Dialogue and Learning in Mathematics Education* (together with Helle Alrø), *Travelling Through Education, In Doubt, An Invitation to Critical Mathematics Education, Foregrounds, Critique as Uncertainty*, and *Connecting Humans with Equations* (together with Ole Ravn). He completed his PhD in 1982 at the Royal Danish School of Educational Studies (mathematics education), and his Dr.Scient in 1995 at Aalborg University.
Orcid: https://orcid.org/0000-0002-1528-796X
Curriculum: http://lattes.cnpq.br/5614296363281466
Email: osk@hum.aau.dk

Daniela Alves Soares is a Ph.D. student in mathematics education at *Universidade Estadual Paulista* (Unesp). Currently, she is a visiting lecturer

during her Ph.D. Exchange at *Freie Universität Berlin*, and a teacher at *Instituto Federal de São Paulo* (IFSP). She has degrees in mathematics and pedagogy. She has worked in teacher education at several institutions, and she has participated in a Brazilian project in teacher education in East Timor. She has worked as a teacher in secondary school, high school, and undergraduate courses. She has published work on critical mathematics education, mathematics for social justice, philosophy of mathematics, and teacher education.
Orcid: https://orcid.org/0000-0002-4527-1184
Curriculum: http://lattes.cnpq.br/4286423748874799
Email: bemdani@gmail.com

Lessandra Marcelly Sousa da Silva has a Master's and a PhD in mathematics education from the Universidade Estadual Paulista (Unesp), Campus Rio Claro, São Paulo, Brazil. She has been a mathematics teacher in public secondary school in São Paulo State since 2004. Her research is on the use of manipulative materials for the teaching of mathematics with blind students, and assistiv e technology resources under the perspective of universal design.
Curriculum: http://lattes.cnpq.br/4241025619581262
Email: lessandramarcelly@gmail.com

Débora Vieira de Souza-Carneiro holds a Master's degree in science and mathematics education from Instituto Federal de Educação, Ciência e Tecnologia de São Paulo. Currently, she is a PhD student in mathematics education at Universidade Estadual Paulista (Unesp), Campus Rio Claro, São Paulo. Her research interests include critical mathematics education, mathematics education in higher education, and problem-based learning. Her teaching experience began in 2005 in public and private schools. Since 2016, she has been teaching higher education in a private teaching institution.
Orcid: https://orcid.org/0000-0002-5137-0883
Curriculum: http://lattes.cnpq.br/2541357847304591
Email: mat_debora@yahoo.com.br

Rafaela Nascimento da Silva has a degree in mathematics from Universidade Federal de Alfenas, Brazil. Her research interests include teaching and learning processes, mathematics education, and the education of seniors. She works as a high school teacher in mathematics.

Orcid https://orcid.org/0000-0003-2880-9437
Curriculum: http://lattes.cnpq.br/9040978472591610
Email: rafaelansil@gmail.com

Jeimy Marcela Cortés Suaréz is from Colombia. In 2020, she obtained a Master's degree in mathematics education from the Universidade Estadual Paulista, Campus Rio Claro, São Paulo. She is now a PhD student in mathematics educations at the same university. The aim of her research is to understand students' subjectivity and conception of social justice. In 2015, she obtained a degree in mathematics from the Universidad Distrital Francisco José de Caldas, Bogotá, Colombia. From 2014 to 2017, she worked as primary school teacher. She is a member of the Creative Imago research group, which explores connections between art and education.

Orcid: https://orcid.org/0000-0003-3894-566X
Curriculum: http://lattes.cnpq.br/6424717833449361
Email: yeyemarch@gmail.com

List of Figures

Chapter 2

1	Registration in the work folders of the students (Gutiérrez and Rodríguez, 2015).	32
2	Translation of the work folder.	33

Chapter 3

1	SRG research classroom in one of the schools.	40
2	Data produced by one of the subgroups in the 2006 edition of the project about the degree of visibility of white and black characters.	48
3	Data produced by one of the subgroups in the 2006 edition of the project, about the quality of the participation of the white and black ethnic groups.	48
4	Subgroup report on the 2010 version of this project.	49
5	Report of the subgroup on the 2010 version of this project.	50

Chapter 13

1	Dani's table. Source: researchers' personal collection.	215

Chapter 14

1	Images showed to the seniors. Source: the internet.	229
2	Images showed to the seniors. Source: the internet.	230
3	Examples of figures presented to the seniors. Source: the internet.	231

4	Examples of figures presented to the seniors. Source: the internet.	233
5	Setting up the table. Source: authors' collection, 2018.	235
6	Eighth and ninth month of the table. Source: authors' collection, 2018.	236
7	Ninth and tenth month of the table. Source: authors' collection, 2018.	237
8	Seniors taking measurements. Source: authors' collection, 2018.	238
9	Golden rectangle. Source: authors' collection, 2018.	238
10	Table completed with the number of adult couples and young couples of rabbits and the total of couples of rabbits for each month. Source: authors' collection, 2018.	239
11	Graph of the relationship between the Fibonacci sequence and the divisions. Source: Belini (2015, p. 39).	240

Chapter 17

| 1 | Participants performing the activity. | 283 |
| 2 | A participant explaining how their group arrived at the result. | 284 |

Chapter 18

1	Structuring elements of the curriculum component Scientific Initiation. High school, IFC, Rio do Sul campus, 2001–2019.	300
2	Adapted from Battisti et al. (2007).	305
3	Taken from Battisti, Campos and Souza (2007).	306
4	Taken from Battisti, Campos and Souza (2007).	306
5	Evolution of windrow height for composting. Taken from Battisti, Campos and Souza (2007).	308

List of Tables

Chapter 1

1	Learning milieus.	4

Chapter 3

1	Population according to colour/race in the neighbourhoods in which the Social Research Group (SRG) schools are located. Data from IBGE Demographic Census 2010. Copyright 2010 by IBGE.	42
2	About the visibility degree. Made by the author.	51
3	The visibility of white characters in the collection made by the students. Table constructed by the author.	52
4	Compiled results from the research subgroups of the SRG in 2017: Groups from ninth grade.	53
5	Compiled results from the research subgroups of the SRG in 2018.	53

Chapter 5

1	Distinctions Between Out-of-School Mathematics and School Mathematics.	79

Chapter 7

1	Content and themes of end-of-unit projects (EUPs).	121
2	Key differences between CME and neoliberal pedagogy.	121

Chapter 14

1	Seniors participating in activities.	225
2	Learning environments. Adapted from Skovsmose (2000).	226

Chapter 17

1	Progressive Table for Calculation of the Monthly Personal Income Tax to Be Paid. Reproduced from Brasil (2017).	280
2	Income Tax Calculation for a Taxable Amount of R$4,000.00.	280

Chapter 18

1	Behaviour of windrows with different residues for composting, Rio do Sul. Taken from Battisti et al. (2007).	307

Index

Name Index

Akkari, A. 138, 141
Almeida, A. C. 278
Alrø, H. 8, 60, 105, 106, 130, 169, 170, 173, 174, 177, 199, 213, 227, 235, 254, 258, 307
Amanti, C. 80
Anderson, G. L. 118
Aranha, A. M. L. 153, 154
Arnon, I. 60
Avcı, B. 116, 117, 121, 123, 125

Barbosa, J. 23, 30, 34
Barros, D. D. 249, 251, 253, 254
Battey, D. 116
Battisti, G. 304, 305, 306, 307, 308
Bazin, M. J. 298
Bazzo, W. A. 296, 297, 299, 317
Beitz, C. R. 136
Belini, M. M. 240
Benevides, M. V. M. 136, 137
Biotto Filho, D. 144
Bishop, A. 186
Blomhøj, M. 23
Borba, M. 305
Bose, A. 70, 71, 74, 78, 82, 84, 86, 88, 89, 91
Britto, R. R. 46, 143, 190

Campana, J. 72
Campos, P. C. 304, 306, 308
Carpenter, T. P. 116
Carraher, D. W. 76, 79, 224

Carraher, T. N. 76
Carrijo, M. 133
Carr, W. 117, 118
Carvalho Jr., P. H. 278
Chapman, O. 60
Christiansen, I. M. 14
Cintra, V. P. 249, 250, 251, 252, 253, 254, 255
Civiero, P. A. G. 297, 298, 299, 301, 302, 308, 311, 317
Clements, D. H. 70

D'Ambrosio, U. 140, 259
Dantas, I. J. M. 143
Descartes, R. 153, 154, 158
Dewey, J. 130
Dijk, T. A. 47
Doll, J. 242
Domingues, P. J. 45

Elkind, E. 189

Fagg, H. 74
Fagnani, E. 278
Farrell, J. P. 124
Faustino, A. C. 8, 105, 144, 167, 171, 180, 181, 277
Fernandez, E. 166
Figueiras, L. 134, 198
Foucault, M. 10
Francisco, J. M. 282
Frango, E. 275, 282
Franke, M. L. 116

Frankenstein, M. 12, 44, 67, 107, 275
Freire, P. x, 10, 11, 12, 15, 41, 73, 104, 107, 110, 125, 145, 150, 151, 160, 169, 173, 227, 253, 255, 259, 277, 287, 297, 301, 304, 305, 311
Freudenthal, H. 74
Fronza, K. R. K. 317

Gadotti, M. 133
Galileo, G. 163, 164
Gandhi, M. K. 72, 74, 75
Gaviolli, I. B. 211, 214, 216, 218
Gearhart, M. 71
Gelman, R. 71
Giroux, H. A. 156
Goldsmith, S. 192
Gomes, M. P. 136
Goos, M. 115
Graven, M. 70
Greenberg, J. 80
Greer, B. 169
Grossi, F. C. D. P. 244
Guarinello, N. L. 135, 137
Guberman, S. R. 71
Gutstein, E. 12, 110, 116, 144, 273, 274, 277, 278, 286, 287, 291, 316

Habermas, J. 118
Healy, L. 198
Herr, K. 118
Heywood, A. 136
Hilgenheger, N. 154

Johnson, D. W. 116, 119
Johnson, F. P. 116, 119
Julio, R. S. 224, 234

Kantha, V. K. 70, 91
Kant, I. 9, 10
Kato, L. 35, 36
Kazemi, E. 116
Kemmis, S. 117, 118
Kollosche, D. 185

Lakatos, I. 10, 178
Larroyo, F. 154
Lave, J. 76, 77, 78, 79

Lawrence, G. 124
Lehrer, R. 116
Lerman, S. 23
Levinas, E. 152, 160
Lima, A. S. 224, 241, 244
Lima, L. F. 13, 224, 225, 228, 234, 241, 244
Lívio, M. 232
Locke, J. 153, 154, 157

Maher, C. A. 282
Maleq, K. 141
Marcone, R. 212, 220
Mariconda, P. R. 174
Marx, K. 9, 10
Massey, C. M. 71
Mehan, H. 170
Milani, R. 105, 106
Moll, L. C. 80, 81
Moreno, Y. C. R. (formerly Rodríguez) 15, 21, 25
Moscovici, S. 224
Moura, A. Q. 13, 144, 167, 186, 187, 202, 277
Muzinatti, J. L. xi, 12, 13

Neff, D. 80
Nielsen, L. 59

Obraztsova, S. 189
Oliveira, F. P. Z. 298, 299, 302, 317
Oliveira, V. C. A. 234
Onwuegbuzie, A. J. 63
Orwell, G. 138

Pais, A. 139, 140
Penteado, M. G. ix, 104, 111, 144, 186, 202, 241, 254
Pietsch, J. 115
Pine, G. J. 117
Plato 154, 165, 166
Postman, N. 314
Powell, A. B. 44, 282

Raymond, D. 249
Reaño, N. 67
Resnick, L. B. 76, 77, 79

Ribeiro, R. D. 282
Rodríguez, F. A. G. (formerly Gutiérrez) 22, 23, 24, 25, 27, 29, 31, 32, 33, 34, 35
Rodríguez, Y. C. R. 328
Rogoff, B. 71, 76

Saldaña, J. 119
Sant'Ana, M. F. 317
Santiago, M. 138
Sarama, J. 70
Saxe, G. B. 71, 76, 79, 80
Scagion, M. P. 224, 244
Scheller, M. 300
Schimidt, F. 282
Schliemann, A. D. 76, 78
Schwab, K. 315
Schwertl, S. L. 317
Scortegagna, P. A. 224
Scribner, S. 76
Seyferth, G. 45
Shakespeare, W. 296
Silva, C. 35, 36
Silva, G. H. G. 224, 241
Silva Júnior, C. A. 249
Skidmore, T. E. 44

Skovsmose, O. ix, 3, 14, 16, 17, 18, 20, 23, 28, 37, 40, 46, 54, 57, 59, 60, 61, 71, 78, 81, 96, 104, 105, 106, 108, 109, 110, 112, 116, 117, 127, 128, 130, 140, 142, 151, 156, 158, 159, 160, 163, 164, 165, 166, 167, 168, 169, 170, 172, 173, 174, 175, 177, 182, 183, 185, 189, 198, 199, 200, 201, 213, 225, 226, 227, 228, 233, 235, 237, 241, 243, 253, 254, 257, 258, 259, 260, 261, 262, 265, 268, 270, 271, 273, 274, 276, 289, 297, 303, 305, 308, 309, 311, 314, 315, 316, 317, 319, 331
Soares, D. A. 150
Socrates 163
Souza, W. O. 304, 306
Steiner, H. G. 189
Subramaniam, K. 82

Tardif, M. 248, 249, 254

Valero, P. 23, 70, 104, 108, 169
Vithal, R. 14
Vygotsky, L. S. 125

Winter, R. 118
Wright, D. E. 116

Zabala, A. 252

Subject Index

absolutism 105, 111, 112, 213, 254, 258
action research 115, 117, 118, 119, 131
autism spectrum disorder xii

Brazilian Institute of Geography and Statistics (IBGE) 337

citizenship xi, 21, 23, 35, 46, 59, 113, 115, 116, 117, 122, 123, 124, 125, 127, 129, 133, 134, 135, 136, 137, 138, 139, 140, 141, 142, 143, 144, 145, 316, 317
civilising equation 295, 296, 297, 298, 301, 315, 318, 319
collaborative learning xi, 115, 116, 117, 119, 120, 121, 122, 123, 124, 125, 127, 128, 129, 130, 131

comfort zone 28, 104, 144, 276
compliance 122, 124, 289
contemporary variable 295, 296, 297, 298, 304, 311, 317
contentious landscapes of investigation 15
contested concept 133, 135, 138, 145
contested construction 3
crisis 121, 296
critical mathematics education ix, x, xi, xii, 1, 2, 3, 11, 12, 13, 14, 15, 23, 39, 44, 57, 58, 59, 60, 67, 115, 116, 117, 119, 121, 123, 124, 125, 126, 128, 129, 130, 131, 142, 164, 186, 193, 257, 258,

260, 261, 264, 270, 271, 273, 276, 277, 278, 291, 292, 295, 297, 298, 302, 318
critical reflection 35, 46, 143, 170, 193, 299, 300, 302, 303
critique 3, 9, 10, 11, 12, 14, 115, 164, 185, 186, 193

deaf students 167, 185, 187, 192, 193, 201, 204, 205, 206, 207, 208, 209
deficiencialism 211, 212, 214, 219, 220, 221
democracy xi, 32, 39, 41, 43, 44, 46, 49, 53, 54, 55, 95, 96, 97, 99, 102, 103, 104, 106, 107, 108, 109, 110, 111, 112, 113, 115, 116, 117, 119, 122, 124, 126, 129, 130, 134, 136, 137, 138, 164, 169, 170, 176, 182, 185, 188, 189, 190, 191, 194, 301
mathematical democracy measures 53, 54, 55
dialogic act 8, 9, 106, 177, 178, 179, 180, 182, 183, 227, 259
dialogue xi, 1, 3, 6, 7, 8, 9, 10, 11, 15, 24, 35, 60, 64, 95, 96, 97, 102, 103, 105, 106, 107, 108, 113, 130, 135, 153, 155, 158, 163, 164, 165, 166, 167, 168, 169, 170, 172, 173, 174, 175, 176, 177, 178, 179, 180, 181, 182, 183, 191, 193, 194, 199, 200, 201, 202, 207, 209, 211, 218, 227, 239, 244, 253, 255, 282, 284, 285, 289, 290, 298, 308, 311, 312, 316
across differences 185, 191, 193, 194
dreams xi, 26, 87, 133, 149, 151, 152, 155, 157, 158, 159, 160

economic and political reflections x, xii, 9, 10, 11, 13, 14, 15, 16, 21, 23, 41, 57, 59, 60, 61, 64, 65, 72, 96, 100, 102, 104, 105, 107, 108, 109, 111, 113, 124, 125, 134, 135, 136, 137, 138, 140, 141, 151, 158, 160, 169, 186, 190, 192, 193, 194, 259, 264, 270, 274, 277, 278, 296, 311
elementary school 143, 163, 168, 170, 182, 197, 198, 211
equality 87, 116, 141
erosion of democracy 55, 96, 185, 188, 189, 190, 191, 192, 193, 194

ethical and moral reflections 58, 59, 61, 64, 107, 108, 120, 140, 154, 158, 190, 264, 270
Exame Nacional do Ensino Médio (ENEM) 263, 264
exclusion 5, 8, 9, 136, 137, 144, 198
exemplary landscapes of investigation 15
exercise paradigm 3, 4, 6, 7, 9, 14, 59, 61, 66, 71, 72, 90, 91, 117, 128, 142, 156, 158, 199, 213, 226, 227, 276, 291, 304, 305, 311
exploitation 11, 12, 88, 89, 135, 136

fairness 69, 72, 87, 90, 96, 160, 189, 282, 284, 286, 291
Fibonacci sequence 224, 228, 231, 232, 235, 239, 240, 242, 244
financial education 21, 37
foreground 21, 23, 37, 71, 149, 151, 152, 155, 158, 159, 160
funds of knowledge 69, 70, 71, 80, 81, 82

global citizenship xi, 133, 134, 141, 142, 143, 144, 145
globalisation 133, 134, 135, 136, 139, 140, 142, 252, 296, 297, 304
golden ratio 224, 228, 231, 232, 234, 238, 240, 241, 242, 243, 244

high school 18, 40, 97, 99, 225, 263, 295, 297, 299, 304, 323, 324, 327, 332, 336

ideology 44, 45, 305
inclusive education ix, 13, 15, 188, 192, 197, 198, 200, 209, 211, 248, 249, 252, 255
inclusive landscapes of investigation xii, 15, 185, 186, 191, 192, 193, 194, 209
income tax 27, 273, 274, 279, 281, 282, 283, 286, 287, 288, 289, 290, 291
injustice x, xi, 11, 12, 15, 16, 60, 69, 72, 88, 89, 137, 140, 145, 277, 316
inquiry 2, 7, 59, 62, 115, 118, 127, 191, 227
inquiry-cooperation model 227
interview 100, 101, 149, 150, 224, 252, 254, 270, 277

investigative activity 156, 257, 258, 259, 260, 261, 262, 264, 265, 266, 267, 268, 269, 270, 271

justice xi, 12, 16, 71, 90, 91, 142, 145, 273, 274, 275, 277, 278, 279, 285, 290, 291, 292, 318, 319

landscape of investigation ix, xi, xii, 1, 2, 3, 4, 5, 9, 13, 14, 16, 21, 23, 27, 28, 30, 31, 34, 35, 36, 37, 39, 46, 49, 58, 59, 60, 61, 62, 63, 65, 66, 67, 71, 72, 77, 87, 89, 90, 91, 105, 106, 109, 110, 111, 127, 128, 133, 142, 143, 144, 145, 159, 160, 163, 166, 167, 168, 188, 191, 192, 199, 201, 205, 211, 213, 219, 220, 221, 226, 228, 231, 233, 234, 235, 237, 238, 239, 240, 241, 242, 243, 252, 253, 259, 261, 266, 269, 275, 276, 282, 289, 290, 291, 295, 298, 302, 303, 305, 318

learning milieu 3, 5, 11, 14, 23, 28, 30, 32, 37, 191, 258, 318

Libras 187, 188, 193, 198, 199, 201, 202, 203, 204, 205, 206, 207, 208, 209, 248, 251, 252, 253

life-world 23, 120, 121, 124, 125, 259

marijuana legalisation xi, 57, 58, 59, 60, 61, 63, 64, 65, 66

mathemacy 40, 88, 95, 96, 104, 105, 110, 257, 259, 260, 270

mathematical and technical reflections 34, 64, 104, 118, 156, 157, 158, 299

mathematical exercise 4, 213, 267

mathematical modelling 21, 23, 34, 36

socio-critical perspective 21, 35, 36

mathematical variable 30, 295, 296, 297, 298, 301, 310, 311, 312, 313, 314, 317, 318

mathematics education for social justice 17, 273, 274, 275, 277, 278, 285, 291, 292

mathematics in action 78, 104, 260, 315

mathematics teacher education x, xii, 14, 247, 249, 255, 260

media 39, 40, 45, 46, 49, 51, 55, 143, 144, 190

meetings amongst differences xii, 197, 198

methodology 16, 117, 118, 211, 250, 251, 258, 299, 300, 305, 311, 319

neoliberal hegemony 117

neoliberal pedagogy 121, 130, 131

non-dialogic act 8, 9, 180, 181, 182

opening an exercise 257, 261, 265, 271

oppression 11, 12, 15, 90, 91, 104, 112, 126, 138, 144, 198, 277

out-of-school mathematics 72, 76, 78, 80

parameter 4, 261, 312, 313

participatory action research 115, 117, 131

polygon 185, 187, 188, 193, 201, 202, 206, 207, 209

professional and technological education 139, 247, 298, 301, 304, 317

quilombola community 95, 96, 100, 107, 108, 109, 110

racism xi, 5, 12, 39, 40, 41, 43, 44, 46, 47, 49, 54, 55, 143, 145

rationality 118, 149, 155, 156, 157, 158, 299, 302

real-life reference 59, 319

reflection 60, 67

Revolutionary Armed Forces of Colombia (FARC) 102, 103, 110

risk zone 104, 111, 144, 241, 254, 276, 290, 305

"sandwich" pattern of communication 213

school mathematics tradition 3, 7, 253, 257, 258, 260, 270

scientific initiation 301, 318

semi-reality 3, 4, 5, 59, 159, 227, 257, 258, 261, 262, 263, 264, 265, 266, 267, 268, 269, 271, 276, 310, 311, 319

senior citizens' education 1, 12, 13, 14, 80, 223, 224, 225, 226, 227, 228, 229, 231, 232, 233, 234, 235, 236, 237, 238, 239, 240, 241, 242, 243, 244, 335

sexism 5, 12
slavery 44, 136, 166, 189
social justice 116, 118, 121, 127, 129, 165, 197
social research group 39, 42, 46, 99
social risk 1, 12, 60
sociology of childhood 75
statistics 16, 41, 65, 67, 143, 190
students in comfortable positions 1, 12, 58, 60
students with disabilities 1, 12, 13, 15, 40, 185, 186, 191, 192, 200, 209, 212, 220, 248, 249, 250, 251

teacher education x, xii, 247, 248, 249, 251, 254, 255, 260, 273, 274, 275, 281, 286, 287, 291
teaching mathematics for social justice 274, 277, 278, 291, 292

technical rationality 149, 156, 158, 299, 302
textbook exercise 267

unfinishedness xi, 149, 150, 151, 152, 158, 160, 173
universal design 185, 192
university students x, 1, 12, 189

visibility 39, 45, 51, 52, 54, 55, 143, 190
degree 51, 52

whitening ideology 44
work context 70, 71, 72, 74, 75, 76, 78, 81, 82

zone of proximal development 122, 125, 126

About the Team

Alessandra Tosi was the managing editor for this book.

Rosalyn Sword performed the copy-editing, proofreading and indexing.

Anna Gatti designed the cover. The cover was produced in InDesign using the Fontin font.

Melissa Purkiss and Luca Baffa typeset the book in InDesign and produced the paperback and hardback editions. The text font is Tex Gyre Pagella; the heading font is Californian FB.

Luca produced the EPUB, AZW3, PDF, HTML, and XML editions — the conversion is performed with open source software such as pandoc (https://pandoc.org/) created by John MacFarlane and other tools freely available on our GitHub page (https://github.com/OpenBookPublishers).

This book need not end here...

Share

All our books — including the one you have just read — are free to access online so that students, researchers and members of the public who can't afford a printed edition will have access to the same ideas. This title will be accessed online by hundreds of readers each month across the globe: why not share the link so that someone you know is one of them?

This book and additional content is available at:

https://doi.org/10.11647/OBP.0316

Customise

Personalise your copy of this book or design new books using OBP and third-party material. Take chapters or whole books from our published list and make a special edition, a new anthology or an illuminating coursepack. Each customised edition will be produced as a paperback and a downloadable PDF.

Find out more at:

https://www.openbookpublishers.com/section/59/1

Follow @OpenBookPublish

Read more at the Open Book Publishers **BLOG**

You may also be interested in:

Advanced Problems in Mathematics
Preparing for University
Stephen Siklos

https://doi.org/10.11647/OBP.0181

The Essence of Mathematics
Through Elementary Problems
Alexandre Borovik and Tony Gardiner

https://doi.org/10.11647/OBP.0168

Teaching Mathematics at Secondary Level
Tony Gardiner

https://doi.org/10.11647/OBP.0071

www.ingramcontent.com/pod-product-compliance
Lightning Source LLC
Chambersburg PA
CBHW041731300426
44115CB00022B/2973